THE WORKS AT BLECHINGLEY TUNNEL.

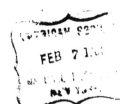

PRACTICAL TUNNELLING;

EXPLAINING IN DETAIL

THE SETTING OUT OF THE WORKS;

SHAFT SINKING AND HEADING DRIVING;

RANGING THE LINES AND LEVELLING UNDER GROUND;

SUB-EXCAVATING, TIMBERING,

AND THE CONSTRUCTION OF THE

BRICKWORK OF TUNNELS;

WITH THE

AMOUNT OF LABOUR REQUIRED FOR, AND THE COST OF THE VARIOUS
PORTIONS OF, THE WORK:

AS EXEMPLIFIED BY THE PARTICULARS OF

BLECHINGLEY AND SALTWOOD TUNNELS.

BY

FREDERICK WALTER SIMMS, F.R.A.S., F.G.S., M.Ins.C.E.

CIVIL ENGINEER,

AUTHOR OF "A TREATISE ON THE PRINCIPLES AND PRACTICE OF LEVELLING," "A TREATISE ON THE
PRINCIPAL MATHEMATICAL AND DRAWING INSTRUMENTS EMPLOYED BY THE ENGINEER,
ARCHITECT AND SURVEYOR," ETC., ETC.

Second Edition,

REVISED, WITH ADDITIONAL PLATES,

BY

W. DAVIS HASKOLL, Civil Engineer,

AUTHOR OF "RAILWAY CONSTRUCTION AND PRACTICE OF ENGINEERING FIELD WORK."

LONDON:

LOCKWOOD & CO., STATIONERS' HALL COURT;

TROUGHTON & SIMMS, 138, FLEET STREET.

1860.

PRACTICAL TUNNELLING.

CHAPTER I.

GEOLOGICAL FEATURES OF THE SOUTH—EASTERN RAILWAY.—GENERAL
ACCOUNT OF BLECHINGLEY AND SALTWOOD TUNNELS, THEIR COST, ETC.

The Blechingley and Saltwood Tunnels are situated upon the line of
the South-Eastern Railway between London and Dover, which
passes through a district of country, not only celebrated for the
beauty of its landscape, but highly interesting to the Geological
enquirer; the Railway being formed through the Tertiary strata,
and the cretaceous group of the secondary formation. Commencing
at the Metropolis, it is constructed upon the London Clay till it
reaches New Cross; where, at about one hundred yards to the
south of the public road bridge, the Plastic Clay formation appears
on the slopes near the bottom of the excavation, in situ beneath

the London Clay. The junction of the two formations at this place is described in an interesting paper read before the Geological Society, April 17, 1844, by H. Warburton, esquire, the President, containing some results of an examination of that locality, in which I had the pleasure of assisting him; and is represented in the following section.

London Clay :

1. { Yellow clay	thickness not determined.	
{ Blue or slate-coloured clay	thickness 10 to 15 feet.	

Plastic Clay formation :

		ft.	in.
2.	Rolled flint pebbles or shingle ... ... thickness	1	10
3.	Fine fawn-coloured sand	0	3
4.	Lignite	0	0½
5.	Fine fawn-coloured sand	2	0
6.	Ferruginous sand, with marine fossils, oyster shells, and cerithia	0	4
7.	Loose grey sand, with fragments of cerithia	0	8
8.	Strong black clay	0	10
9.	Black clay and sand, with fragments of oysters and cerithia ...	0	9
10.	Black dirty sand	0	4
11.	Dark sand, containing fossils, oyster shells, &c.	0	6
12.	Calcareous stone, containing paludina, unio, &c. (freshwater fossils)	0	6
13.	Decomposed stone and sand, with oysters, &c.	0	3

The shells belonging to the upper part of the Plastic Clay series in this neighbourhood have been well described by Dr. Buckland in the fourth volume of the first series of the Geological Society's Transactions, but the occurrence of the paludina and unio in the stratum No. 12 of the above section, which are freshwater shells, thus included between marine fossils, appears to have escaped observation, till now discovered by Mr. Warburton; who describes the stone in which they are embedded, as septaria of a texture considerably more earthy than the septaria of the London Clay usually are.

The line of Railway continues upon the Plastic Clay as far as Combe Lane, Croydon, where the Chalk crops out from beneath the sands of the last named formation, and is distinctly to be seen on the north-east slope of the cutting. The Railway then crosses the great Chalk range that extends from Dover to Hampshire, and rises towards the Chalk escarpment at Merstham in Surrey, where its greatest summit level between London and Dover is attained in the tunnel near that place.

In the deep cutting at the south of the tunnel, a good section of the Upper Green Sand stratum appears, cropping out from beneath the Chalk; this is succeeded at the village of Merstham by the Gault, through which the road to Blechingley has been lowered that it might be passed under the Railway.

At a short distance further southward, the Lower Green Sand formation rises to the surface, in beds of fawn-coloured sand, very siliceous, and good for Engineering purposes. The middle beds of the Lower Green Sand, as indicated by the presence of rushes and wet land, next appear; and these are followed by the lower beds of the same formation, which contain the Kentish ragstone, fuller's earth, &c.—the fuller's earth pits of Nutfield being near this locality. The lower beds of this formation rise to a considerable height, and form the range of sand hills that passes through the country, parallel to the great chalk range before named.

The place where the railway crosses the sand range is called Redstone Hill, and is the point where the Brighton railway diverges to the south, while the Dover railway passes round the hill with a curve of half-a-mile radius to the eastward; and towards the further end of this curve, near to a bridge at Robert's Hole farm, the next inferior stratum, the Weald Clay, emerges from beneath the sand. This spot may be further identified by the greater width of the excavation, or flatness of the slopes, occasioned by the slipping of the earth at the junction of the two formations, where much water was present. In making this excavation, some stone was found, that was much jointed, and contained innumerable fossils, which, upon examination in April 1843, by Mr. Warburton, Dr. Fitton, Mr. Austen, and myself, was found to include some of the most characteristic of M. Leymerie's Neocomien species, with a few belonging also to the quarrystone of Hythe; as, arca raulini, panopœa depressa, pholadomya acutisulcata (Leymerie), pecten obliquus (interstriatus), pinna sulcifera, gervillia aviculoides, perna mulleti, p. alœformis, trigonia dœdalea, t. Fittoni, gryphæa sinuata, nautilus radiatus, &c. This stone appears to correspond with the Atherfield rocks in the Isle of Wight; which it resembles in its mineralogical and geological character. [See paper by Dr. Fitton, read before the Geological Society, May 24, 1843, entitled "Observations on the Section of the Lower Green Sand at Atherfield, on the coast of the Isle of Wight."]

From Redstone Hill the line passes eastward, along the Weald Clay, in successive cuttings and embankments for many miles, except that near Blechingley there is a tunnel, bearing that name, formed through a spur of Tilburstow Hill: the Weald Clay at this place is indurated into a shale, or blue bind, and being full of joints and faults, caused much difficulty in the work, as will be described in the following pages. The fossils found during the construction of the tunnel, were portions of the iguanodon, hylœosaurus, cypris, paludina, clathraria (Lyelli), &c. &c., and a fine specimen of the lepidotus (Mantelli), presented by me to the Geological Society, accompanied by a short paper upon the subject of the strata at this place, which was read at the Society's meeting, on February 21st, 1844.

Near the town of Ashford the line leaves the Weald Clay, and again enters upon the Lower Green Sand formation, which continues to be its base as far as Folkestone, a distance of about fifteen miles; the summit is passed by means of a Tunnel, at Saltwood, not far from the out-crop of the Sand from beneath the Gault; consequently the shafts were sunk through the upper beds, and the tunnel is formed at the junction between that and the middle bed; where a large quantity of water was encountered, which greatly retarded the progress of the works. Among numerous fossil remains found at Saltwood, chiefly in ferruginous concretions, the following may be particularly enumerated: nautilus radiatus, gervilia aviculoides, terebratula, tethys major, panopœa, trigonia alœformis, venus, cardium, tornatella, pecten quinquecostatus, p. orbicularis, &c. &c., with fossil coniferous wood pierced by gastrochæna; together with a remarkable product, a new and beautiful resin, which partakes of the properties of amber and of retin-asphalt, and is principally marked by its clear red colour, its infusibility, and the difficulty with which it is acted upon by many chemical solvents. I was indebted to Mr. Edward Solly, through the kindness of Dr. Fitton, for a chemical examination of this substance, the results of which are inserted at length in a paper read before the Geological Society, June 7th, 1843, giving an account of an investigation of the strata from the summit of the Chalk escarpment above Saltwood tunnel to the sea at Hythe; or, at right angles both to the range of hills and the direction of the line of Railway in that locality. It may not be uninteresting to insert the result of such examination.

The upper Green Sand stratum, which at the back of the Isle of Wight is one hundred and four feet thick, is here altogether wanting, it having thinned out at this place.

Strata from beneath the Chalk to the Wealden, through Saltwood.

		ft.	in.
Upper Green Sand, ... (wanting,—but) at Folkestone, five miles distant, it is in thickness		15	0
Gault		126	0
Lower Green Sand.			
Upper Division	70 0		
Middle ditto	158 0		
Sand above the Quarries 67 0			
Quarry Rock 48 0			
Sand and Stone, previously concealed 14 0	... 178 6	406	6
Clay beneath the sand and stone ... 49 6			
Total thickness from the Chalk to the Wealden		547	6

Near Folkestone station the line leaves the Sand, and crosses the Gault formation; where, at the junction of that stratum with the Upper Green Sand, and then of the Chalk above, a tunnel is made through the hill to the undercliff called the Warren, and from thence to Dover, entirely in the Chalk, through and along the face of the cliffs;—altogether one of the grandest Engineering Works in the kingdom;—and where Mr. Cubitt, the Engineer-in-Chief to the Railway Company, so successfully introduced the use of gunpowder

in blasting rock upon a great scale, especially in removing that
large mass of chalk, " The Round Down," on January 26th, 1843:
the particulars of which are given by Lieutenant Hutchinson, R. E.,
in the sixth volume of the Professional Papers of the Corps of Royal
Engineers.

Such are the Geological Features of this line of Railway.

The Engineer-in-Chief charged with the construction of the line
was William Cubitt, Esq., F.R.S. &c. &c.   That gentleman divided
the whole line into three districts; over each of which he appointed
a resident Engineer.   To that nearest London he appointed the
author; the district through the Weald of Kent was assigned to Mr.
Barlow; and the third, or Dover district, was given to Mr. Wright.
In the district first named the Blechingley Tunnel is situated: and,
upon its completion, the author was further appointed to superintend
the construction of the Tunnel at Saltwood :—the particulars of which
two works form the subject of the following pages.   And to my
colleagues, and myself, it is gratifying to know that our Engineer-
in-chief has expressed himself, both in public and private, satisfied
with the manner in which we, severally, have carried his views and
intentions into execution.

A general description of the Blechingley and Saltwood Tunnels,
explaining their cost, and the circumstances under which they were
constructed, is annexed, previously to entering upon the details of
the same.

BLECHINGLEY TUNNEL.

So named from the Parish where it is situated, is in the county
of Surrey, and about twenty-five miles from London.   The Tunnel
is twenty-four feet wide in the clear, and twenty-one feet from the
upper surface of the rails to the crown of the arch; its figure is
elliptical above the skewback, or springing of the invert; the versed
sine of the invert is three feet, and the level of the rails is one
foot above the skewback.   The Tunnel is inclined from west to
east, at the rate of three feet per mile.   The dimensions of the brick-

work varied, and were regulated according to the appearances of the ground, from time to time, as the lengths, which were twelve feet long, were excavated: these particulars are given in figure 3, plate I. the left half of which shews half the cross-section of the Tunnel at Blechingley; and the right half, that at Saltwood. Figure 1, plate I., is a longitudinal section of Blechingley Tunnel, and figure 2 that at Saltwood; shewing the positions of the working shafts, and of the Observatory, &c.—all of which subjects will be described in further details in the course of the work in each tunnel.

The ridge through which the Tunnel passes is the main axis of elevation of this part of the country; and, from the dip of the strata in both directions from its summit, forms a north and south anticlinal axis; its direction being that of the meridian, nearly; which, so far as I can judge, extends from the chalk range between Godstone and Merstham in Surrey, to about Ditchling in Sussex: the waters which fall on the surface along the said line of direction, form sources or feeders to the rivers Medway and Ouse, on the east, and to the Mole and Adur on the west. Besides the inclination of the beds both ways from the axis they dip to the north at an angle of about thirteen degrees; but westward, from the summit of the ridge, there is no regularity in this respect, the strata lying as it were in heaps, at almost every angle, from five to sixty degrees, and dipping in all directions, from west-by-north to east; besides which a detached mass of sand-rock lay across my path, near the top of the Tunnel, and from whence a great body of water was discharged into the workings, causing no small trouble and difficulty.

The Blue Clay of the Weald in which I was working was at first greasy to the touch; and when dry, and in situ, formed a hard shale, requiring an extensive use of gunpowder in its excavation, but upon exposure to damp and atmospheric action, it swelled considerably and then slaked: this obliged me to close-pole the face of the work in all directions as far down as the lower sill, and frequently to the bottom. The expansion, or swelling, was occasionally so great as to threaten the hurling in of the lengths after they were completely timbered, and would probably have done so but for constant watchfulness, and strong timbers properly applied. The pressure upon the work was sometimes so great that sound oak bars, fourteen or fifteen inches in diameter, were cracked and broken as if they had been mere sticks.

The pressure we had to contend with was variable, and uncertain in the highest degree; sometimes a length could be got out, and the arch turned, without any apparent movement of the earth around and above us; at other times, the ground when partly excavated would begin to move, and press upon the bars on one side of the arch; at others, it would act upon the crown bars; the former action was principally confined to those parts of the Tunnel that were deepest below the surface, whilst the greatest pressure (which mostly acted upon the crown) took place towards the ends of the Tunnel, where the surface was so much nearer to the arch. It sometimes occurred that after a length had been excavated satisfactorily, and by the time the bricklayers had built up the side walls, the weight on the top was so great as to press the bars down to an extent nearly equal to their own thickness, which was seldom less than fifteen inches; consequently, when the centres were set, there was not sufficient space between them and the bars to insert the brickwork of the arch; the remedy for this was, to remove the poling, and excavate more earth from above the bars, and to prop them again at a higher level; which occasioned considerable loss of time, and consequently increased the expense.

In one of the pits—No. 11, at the east end,—this weighting of the crown occurred constantly after getting in three lengths west of the shaft, and therefore we elevated the extremity of the bars sufficiently high, upon their first insertion, to allow for the expected subsidence: in the other pits its occurrence was uncertain; consequently it was impossible to provide for it, without running the risk of having a great opening above the arch, to be filled solid with brickwork, or with earth, which is often imperfectly done, and would be liable to bring a greater weight upon the work when the earth again takes its bearing, after the mass has been in motion. It is the general movement of the mass in adjusting itself to equilibrium, after the disturbance occasioned by the excavation, that causes the weight, and whose searching influence finds out the weak points in the work.

The greater pressure upon the work in shallow ground over that where the tunnel is very deep below the surface, I can explain only upon the supposition that in the former case, the whole superincumbent mass is acting perpendicularly downwards; whilst, in the latter case, a small portion only gets into motion, the upper part acting as a key, (if I may so express myself,) by which the mass supports itself. This action was clearly shewn in pit No. 11, above referred to; where the working below could be distinctly traced upon the surface of the ground, by its sinking in the form of a basin as our work proceeded, and at the same time cracking into large fissures.

The sinking of the shafts was commenced in the beginning of August, 1840. These were down to the depth necessary for the shaft-sills by the middle of September; which, together with the further sinking, including the square timbering to the bottom of the Tunnel, was completed by the end of October. The driving of the heading at the level of the top of the invert was then commenced, and was finished by Christmas. From this time till February 12th, 1841, preparations were made for commencing the excavation of the Tunnel: these consisted in making a gin for each shaft, and the ground-moulds, leading-frames, and centres. On the above date the miners broke ground in No. 3 pit, being the first commencement of the tunnelling; but it was not until early in April that the whole of the shafts were got to work; and as soon as each was started, the work was pushed on with the utmost vigour, night and day.

On September 3rd, the first junction was effected; and on November 1st, the last junction was keyed in; the Tunnel, as originally intended, was therefore complete; but it was resolved to extend it at each end in consequence of the backwardness of the open cuttings that were let to two different contractors. My instructions were, to extend the tunnel until I should meet the open cuttings, and thus enable the Directors to open the Railway to the public at the time proposed, which otherwise could not have been done. The extension of the Tunnel, and the erection of the entrances, were not completed until early in the following May;

c

and the Railway was opened to the public on May 26th. The Tunnel, as completed, is 1324 yards in length.

Although it would appear, from what has above been stated, that the sinking of the shafts did not commence until the month of August,—yet this must be understood as in reference to the working shafts only; because, in the preceding February, two trial-shafts were sunk, to ascertain the character of the ground in which the Tunnel was ultimately to be constructed. The particulars of this work will be given in Chapter IV. After two trial-shafts had been sunk, it still appeared desirable to examine the ground at two intermediate points; accordingly two other shafts were commenced early in the spring, and, to save expense, they were made the full size of working shafts in the first instance, with the intention of employing them as such in the course of the work. The working shafts were 9 feet diameter in the clear, while the trial-shafts were but 6 feet. These large shafts had, however, not been far proceeded with, when an unpleasant difference arose between the Company and the Occupiers of the land, who demanded an exorbitant amount of compensation, forbidding the proceedings until such was paid. Under these circumstances the work was suspended until the following August; when the said differences having been adjusted, possession of the land was obtained, and the works were prosecuted with vigour.

Previously to laying the permanent way, a culvert was constructed upon the invert, throughout the Tunnel, as shewn in section, Figure 3, Plate 1. The Tunnel was also lime-whited twice, with a view of increasing the light; but this did not answer as was expected.

The monthly rate of progress, during the time the work was in full activity, was as follows. During May, 1841, 104 yards were completed;—June, 185 yards;—July, 264 yards;—and August, 228 yards. The bricks were all made on the ground, and wheeled or carted to the various shafts; their cost when thus delivered at the pit's mouth, including waste and all other expenses incurred, was £2 1s. 6d. per thousand. A portion of the bricks was made during the winter of 1840, and dried in flues, by coal fires; which increased the cost considerably. [See paper by the author, on this subject, read before the Institution of Civil Engineers, April 25th, 1843.]

The following abstract will shew the whole cost of this important work.

ABSTRACT OF THE COST OF BLECHINGLEY TUNNEL.

	£ s. d.	£ s. d.	£ s. d.
MATERIALS :—			
Bricks	30,499 12 10		
Cement	11,016 0 11		
Timber	11,341 19 2		
Wrought and cast ironwork, and ironmongery	2,499 3 1		
Miscellaneous: including pumps, weighing machine, broken stone for roads, lime, ropes, stationery, and all materials not included under any of the above heads	6,555 2 8		
		... ...	61,911 18 8
LABOUR :—			
Mining. { Shafts, heading, and preliminary works	3,273 2 8		
Driving the tunnel —including the hire of gin-horses, and the open excavation, for lengthening the tunnel	15,727 7 0		
		19,000 9 8	
Brickwork. { Shafts, and preliminary works ...	378 8 0		
Constructing the tunnel, and lengthening the tunnel in open excavation	11,265 4 11		
		11,643 12 11	
MISCELLANEOUS :—			
Including the erection of the tunnel entrances, culvert through the tunnel, part ballasting the tunnel, construction of machinery, erection of buildings, carpentry, sawing, clerks and inspectors' wages, &c. &c.		6,980 16 6	
			37,624 19 1
			99,536 17 9
Deduct estimated value of plant, removed to Saltwood, upon the completion of Blechingley Tunnel ...			4,300 0 0
TOTAL COST of Blechingley Tunnel ... £			95,236 17 9

Being at the rate of £71 18s. 7d. per lineal yard, for the whole tunnel; 1,324 yards in length, or three-quarters of a mile and four yards.

The foregoing statement sets forth the total cost of the Tunnel, and the price per yard forward: the details and expense of each portion of the work will be given in the following chapters, under their respective heads.

The construction of the Tunnel was entrusted, by Mr. Cubitt, wholly to the author, who at the same time had charge of the works let to contractors, extending over nearly fourteen miles of the Railway. Upon its completion he proceeded with the plant to Saltwood to construct a similar Tunnel there. In the execution of the preliminary works at Saltwood, difficulties of no ordinary character occurred; but by overcoming them the work was subsequently made comparatively easy, as will hereafter be described. The Board of Directors afterwards determined upon letting the construction of the Tunnel (itself) to contractors; leaving the author to superintend the same, and look after the interests of the Company. Under these circumstances, the cost of each portion of that work cannot be given in detail, as will be done in the case of Blechingley Tunnel, except for the preliminary works; but the total cost will be supplied; and, what will be at least as important, the amount or quantity of labour expended in the construction of various parts of the Tunnel will be given, and a comparison drawn between such statement and the corresponding labour at Blechingley.

SALTWOOD TUNNEL.

This Tunnel is also named after the Parish wherein it is situate; and is 954 yards in length. Its form and dimensions are precisely the same as of that at Blechingley, except that the versed sine of the invert was made three feet six inches, instead of three feet. The thickness of the brickwork did not vary so much, neither was its average thickness so great; which arose from the more homogeneous character of the ground, after it had been drained by the preliminary works. The Tunnel is inclined towards Dover, at the rate of twenty feet per mile, or 1 in 264. An examination of figs. 2 and 3, plate I., will shew all that is requisite of these particulars. The former is a longitudinal section; and the right-hand half of the latter is a cross section of half the Tunnel.

The Tunnel is constructed at the top of the middle beds of the Lower Green Sand; and is about ninety feet below the surface of the ground. This stratum contained, as it usually does, a great body of water, which occasioned the difficulty in sinking the shafts, and driving the heading, that will hereafter be described.

The sinking of the working-shafts, which were nine feet clear diameter, was commenced on the 11th of June, 1842, and was carried on in the usual manner. The earth was raised to the surface by means of the common windlass, or jack-roll, at which four men could work: and had circumstances continued favourable, it was intended that no other machinery should have been used for that purpose, until the sinking and heading were completed; as had been previously done at Blechingley. An oak or elm curb was inserted at the bottom of each excavated length, on which the brickwork was carried up to underpin the preceding curb. The sections of the shaft, plate II. shew the windlass, &c. and the curbs inserted in their places; and give a correct notion of the whole. The bottom curb was so placed that the brickwork of the shaft should terminate eight feet above the level of the intended soffit of the arch.

The difficulties commenced about July 13th; when the ground became a perfect quicksand. The sinking of the shafts, and driving of the heading, henceforth became difficult, tedious, and expensive, compared with what had been expected. Manual labour was insufficient to draw the water to the surface, and the horse-gins that had been brought from Blechingley were erected, and barrels holding one hundred gallons were substituted for the twenty-gallon buckets worked by the men. The average quantity of water drawn from each shaft, during the remainder of July, was 700 gallons per hour, when it increased to 1,600 gallons per hour, which brought with it large portions of sand from the back of the timbering, and added to both the difficulty and danger of the work; to secure the sand from running, the back of the polings was, at the time of their insertion, well packed with straw, which let the water pass, but retained the sand: this kind of packing was subsequently used throughout the work, and in all cases, when properly done, it answered the purpose.

By the end of August the average quantity of water drawn from each of the pits amounted to 23 barrels of 100 gallons each per hour; whilst at the same time no more than one-half of a cubic yard of sand was raised to the surface. On the 31st of that month, the quantity of water drawn from No. 7 pit averaged 37½ barrels per hour, which, together with the small quantity of earth raised, gave, as the work of each horse for three hours, 42,637 lbs. raised one foot high in a minute.

The miners complained much of their employment, and were obliged to have three, and in some cases four shifts in every twenty-four hours; they were wet to the skin in a few minutes after they entered upon their work, and, in many instances, illness succeeded their constant working in the water. In addition to the quantity of water, we were troubled with foul air in shafts No. 10 and 13, which prevented the candles burning. (The latter was a supplementary shaft, to facilitate the bringing up the heading along the intended open cutting, and, therefore, is not shewn in the longitudinal section in plate I.) The headings thus circumstanced were ventilated by an air-machine, or what the workmen commonly call a Blow-George, fixed at the top of the shafts; which is simply a fan-wheel revolving rapidly in a spiral-formed box, driving fresh air into the workings below, through tubes fixed in the shafts and along the headings, where the men were at work.

About the end of September, the person who supplied the horses wished to withdraw them from the works, as so many had been knocked up, and several had died. Accordingly, endeavours were made to obtain others; and notices were issued around the country, advertising for horses; but the assistance thus obtained was very trifling, as the numerous carters then in the neighbourhood declined putting their cattle to the labour. The same party was therefore prevailed upon to continue his help.

The following statement will shew about the amount of labour exacted from each horse, at the end of August, at the shafts then at work.

Number of Pounds raised one foot high per minute, by each Horse.	Number of Hours each Horse worked.
24,475	6
22,626	6
24,535	6
42,637	8
22,584	6
80,820	8

More minute particulars of the difficulties arising from the influx of water, and of the mode of proceeding whereby they were overcome, will be inserted in Chapters VI. and VII.; wherein the square timbering and heading driving are described,—as being a more appropriate place for their insertion.

These exertions were necessary, because it was desirable that the preliminary works should be got through without having recourse to steam power for pumping; as that would at once have involved the Company in a great expense; especially, as it appeared probable that if a heading, or adit, could be made quite through the hill, at the level of the bottom of the tunnel, with an outlet at the east end leading to the natural drainage of the country, the whole of the water would run off from the works, as it collected, and leave them sufficiently dry to admit of the work being proceeded with; which was as much as, at that time, appeared reasonable to expect, the proposed inclination of the tunnel in that direction being ample for the purpose. This, by perseverance, was accomplished; but it was not until the end of October that the heading was completed, and the drainage effected.

The heading, when completed, was altogether 1250 yards in length; as it not only extended throughout the length of the tunnel, but also under the open cutting at the east end, before an outlet could be obtained for the water; which there runs off into the valley of Newington, and joins the natural drainage of a large district of country that empties itself into the sea, at Hythe.

When the heading was effectually opened, it was considered desirable to ascertain the quantity of water that was continually passing away from it. Accordingly, a gauge was fixed at its outlet, which narrowed the opening to twelve inches in width, and the depth of water flowing over the waste board (or between the sides of the twelve-inch opening) was noted, from time to time. At first the depth was five inches; which indicated a discharge of about 359 gallons per minute, or 21,540 gallons per hour; but this quantity afterwards diminished, and for a length of time it averaged four inches in depth, which gave about 257 gallons per minute, or 15,420 gallons per hour. The first of these quantities being noted as soon as the heading was completed, confirmed the accuracy of the registered quantity of water drawn to the surface by horse power, as will be explained in Chapter VII.

The following table shews, approximately, the quantity of water, in gallons, and in cubic feet, per minute, that passes over a waste board twelve inches wide.

Depth on the waste board in inches.	Water, per minute.	
	In Imperial Gallons.	In Cubic Feet.
1	32·1	5·14
2	90·8	14·53
3	167·0	26·71
4	257·0	41·12
5	359·2	57·47
6	472·1	75·54
7	595·0	95·20
8	726·9	116·30
9	867·3	138·77
10	1015·1	162·42

For any other width of gauge, a proportionate quantity of water will be discharged.

On the 1st of November the Board of Directors contracted with Messrs. Grissell and Peto, and Betts and Son, to construct the Tunnel for the sum of £85,000, exclusive of the entrances, they allowing the Company £3,006 for the plant and materials, as they then stood upon the ground. The whole to be completed on or before the 1st day of

June, 1843, or the Contractors to forfeit the sum of £50 per day, from that date.

The work was executed under the author's superintendence, as Resident Engineer, and completed by the time named. The preliminary works that cost so much labour and anxiety during their construction fully answered the purposes for which they were designed, and drained the ground more effectually than was anticipated; this necessarily facilitated the progress of the work, and was highly advantageous to the Contractors.

It now only remains to give an abstract of the cost of this tunnel, before proceeding to the details of the setting out and the construction of the works.

ABSTRACT OF THE COST OF SALTWOOD TUNNEL.

Expenditure previous to the time that the Works were contracted for.

	For Preliminary Works.	For the Tunnel.	SUMS.
	£ s. d.	£ s. d.	£ s. d.
MATERIALS :			
Bricks, including carting ...	3,210 17 3	14,527 12 7	17,738 9 10
Cement	... ...	253 2 6	253 2 6
Lime	50 14 4	... ...	50 14 4
Timber	1,928 9 4	1,420 1 4	3,348 10 8
Ironwork and ironmongery	317 5 5	159 2 2	476 7 7
Straw	119 7 10	... ...	119 7 10
Ropes	86 8 9	... ...	86 8 9
Cottages, office, store, &c.	62 0 0	248 0 0	310 0 0
Roads along the whole of the works	70 0 0	400 18 7	470 18 7
Miscellaneous materials ...	519 2 3	187 17 2	706 19 5
Plant from Blechingley Tunnel	1,433 0 0	2,867 0 0	4,300 0 0
CARTING and HORSE-HIRE :			
Carting on works	251 9 2	103 15 6	355 4 8
Hire of gin-horses ...	1,585 15 3	... ...	1,585 15 3
Materials from Blechingley Tunnel	315 11 9	631 3 5	946 15 2
Freight, &c.	120 0 0	292 18 0	412 18 0
LABOUR :			
Mining	3,477 11 11	... ...	3,477 11 11
Brickwork	217 3 4	... ...	217 3 4
Sawing	67 0 8	... ...	67 0 8
Miscellaneous	706 8 11	682 17 1	1,389 6 0
MISCELLANEOUS EXPENSES ...	70 10 0	... ...	70 10 0
TOTALS ... ... £	14,608 16 2	21,774 8 4	36,383 4 6

D

This amount added to the expenditure under the contract, will give the total cost of the Tunnel,—as follows:—

ABSTRACT OF THE COST OF SALTWOOD TUNNEL.

Expenditure under the Contract, after the Preliminary Works were completed.

	£ s. d.	£ s. d.	£ s. d.
Amount of Contract		... ...	85,000 0 0
Hoop-bond joints		32 10 8	
TUNNEL ENTRANCES:— £ s. d.			
Excavation 75 8 9			
Brickwork 1,553 0 0			
		1,628 8 9	
			1,660 19 5
Additional brickwork		321 15 0	
„ timber		176 8 0	
„ labour		358 8 0	
Hard white bricks, for arch ...		2,948 3 6	
Broken stone for ballast		99 11 6	
Drains, cesspools, &c., at ends of Tunnel		386 5 10	
			4,290 11 10
			90,951 11 3
DEDUCTIONS:—			
Tunnel 10¼ yards short of 964 (the contract length)		825 2 6	
Saving in shafts closed		36 0 0	
Saving upon white brick not carted ...		223 15 7	
Bricks for which the Company had previously paid [see last page] ...		14,527 12 7	
			15,612 11 8
			75,338 19 7
Allow for plant, as per contract			3,006 0 0
			£72,332 19 7

Now, if to the above sum be added, the amount expended previous to letting the contract, as before set forth, and the wages afterwards paid to inspectors, &c. together with £3,006—the assumed value of the plant,—the whole cost of the tunnel will be shewn.

	£ s. d.
Preliminary Works, and previous expenses	36,383 4 6
Payments under contract	72,332 19 7
Inspection, rent of land, sorting bricks, &c., &c. ...	820 1 5
Assumed value of plant	3,006 0 0
Total Cost of Saltwood Tunnel ...	£112,542 5 6

Being at the rate of £118 per lineal yard for the whole Tunnel; 953¾ yards in length, or half-a-mile and 73¾ yards; but upon a very careful admeasurement the Tunnel proved to be very little short of 954 yards.

The bricks for Saltwood tunnel had been contracted for previously to the author's leaving Blechingley; they were made at Folkestone, averaging five miles distant from the works; and the cost when delivered was 51s. per thousand.

The quantity of bricks used in the construction of Blechingley and Saltwood tunnels; including the entrances, culverts, shaft towers, and all contingent works, was as follows:—

Blechingley	...	...	14,696,005	or	11,099 per lineal yard.
Saltwood		...	10,186,246	or	10,677 per lineal yard.

CHAPTER II.

DESCRIPTION OF THE OBSERVATORY.—THE TRANSIT INSTRUMENT, AND METHOD OF FIXING AND ADJUSTING.

TUNNELS have generally been made straight from end to end; and I believe that the few exceptions (or curved tunnels) are of comparatively short lengths, and constructed through good ground, that would admit of a large excavation, or cavern, to be made throughout their whole extent, or nearly so, without requiring much timber to support the strata; consequently there could be but little difficulty in forming such a tunnel to the curve required: but in heavy ground they are made straight, not merely on account of the greater difficulty of construction otherwise (except in peculiar situations), but, for Railway purposes, it would be considered unsafe for the trains to be passing and repassing each other in darkness and a confined space, without the advantage of each engine-fire being visible from the other train for some distance.

The Blechingley and Saltwood Tunnels are both straight, and the centre line was carefully ranged with a transit instrument of thirty inches focal length, having an object-glass of two and three quarters inches aperture mounted on a cast-iron stand. In order to command a view of every shaft on the work, the instrument was set up on the most elevated spot of ground, as near the middle of the tunnel as possible; and, that the view might be uninterrupted by the machinery and timber about the shafts, as well as the earth when brought up from below as the work proceeded, the transit was elevated considerably above the surface, by the erection of an observatory; and as such a building is only required during the construction of the tunnel, it is generally but a temporary erection; although there are instances of observatories for such works having been built of an expensive character. The observatories at Blechingley and Saltwood were nearly similar to each other; and the following engraving, shewing that at Saltwood, will give the reader a knowledge of the kind of building that will be sufficient for all such purposes, and, being composed of brick and timber, it may be taken down at an advanced period of the work, and the materials used up.

THE OBSERVATORY.

The annexed engraving shews a section of the Saltwood Observatory, taken in the direction of the Tunnel, with the brick pier in the centre, surmounted with the Transit Instrument. The pier, which was 30 feet high, was erected over the centre-line of the intended tunnel, and quite independent of the rest of the building; so that any motion given to the building by high winds, or otherwise, might not be communicated to the pier. The object to be attained was great steadiness to the Transit, which could not be ensured so well as by an erection of brick-work. The dimensions of the pier, and its counterforts, are given in the engraving. A flat stone was set on the top of the pier, to which the iron stand of the transit was screwed down, as shewn in detail in the engraving of the instrument itself, which will be given in the next page.

The building requires but little description: it was composed of larch poles, intended for and afterwards used in the works, stiffened with struts and cross-pieces; the upper part, or observatory-room, was enclosed with quartering and feather-edged boards. The ascent was by steps from below through a trap-door in the floor; and the floor was trimmed so as not to touch the brick pier by about 5 or 6 inches. Narrow openings were made in the sides of the room, in the direction of the tunnel, through which the observer might look each way for the ranging of the lines; these were closed with small sliding shutters, so that one or more of them could be opened at a time, leaving either a large or small aperture, as occasion might require. The whole was surmounted by a telegraph, having two arms for the observer to signal for the ranging lines to be moved either to the right or to the left; and when the line was found to be correct, both arms were extended, thus denoting that the line was to be moved both to the right and to the left at the same time,—a thing impossible,—it was therefore to be fixed in its then position. In the engraving (facing the title-page), representing the works above ground at Blechingley, the telegraph is shewn as signaling to the men setting the lines on the ranging-frame of the shaft at the right-hand of the picture.

THE TRANSIT INSTRUMENT.

The annexed engraving represents the Transit Instrument; which consists of a telescope, having an axis at right angles to its length, supported on a cast iron stand. This is a standard instrument in every astronomical observatory; where it is adjusted to describe or define a vertical circle passing from the north to the south points of the horizon, 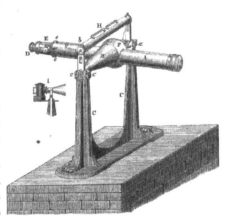 through the zenith of the place and is the best means of observing the passage of the celestial bodies across the plane of the meridian from which *time* is correctly derived: hence its name, "Transit Instrument;" and, thus employed, it may not inappropriately be called the *hand* which points to the time as shewn by that unerring dial, the starry heavens. The same construction which renders it the instrument best adapted to trace a vertical plane for astronomical purposes, makes it equally so to set out a right-line on the surface of the earth; or, our problem more properly is,—*to find any number of points in a straight line connecting two given distant points, the instrument to be situated also between the given distant points.* The line thus connecting the distant points is the base of a vertical plane of small extent. To do this, the telescope, A A', is made to revolve vertically upon a horizontal axis, B B, the pivots of which are supported by the upright arms, C C, of the iron stand.

It does not appear to be necessary in this place to explain how a vertical circle is described by such an instrument, but that its performance may be correct it is essential that the optical axis of the telescope, or line of collimation as it is called, should be precisely at right angles to the horizontal axis about which it revolves; and also that the extremities or pivots of the said horizontal axis where they rest in their bearings on the iron stand be precisely level with each other.

The telescope resembles those of theodolites and spirit levels, and for terrestrial purposes is supplied with a similarly-arranged system of cross wires, the intersection of the centre wires with each other, represents when in proper adjustment, the line of collimation, or optical axis of the telescope. The slide D, or eye-piece, is movable in or out, to obtain distinct vision of the cross wires; an adjustment that must be made or verified each time that the instrument is set up for use, and so long as its eye-piece remains undisturbed it will require no repetition for the same eye; but would, in all probability, require alteration to suit the eye of another person. In adjusting the eye-piece to obtain distinct vision of the wires, it will be found that it can be accomplished with greater certainty by directing the telescope to a white sheet of paper, held or fastened at a little distance off, or even pointed to the sky, so that the wires may appear to be projected on a clear disc.

The screw E gives motion to a rack and pinion, which is the means of diminishing or increasing the distance between the object

and eye-glasses, or rather I should say, between the object glass and
the cross wires, and thereby produces distinct vision of the distant
object to be observed. It should be ascertained that both of the
above adjustments be quite perfect, before any determinate observation
is made, otherwise a parallax will exist, which will prevent an
accurate result being obtained: what is called parallax, in this case,
is an apparent motion between the object viewed and the wires of
the telescope, and is detected when the observer moves his eye up
and down, or sideways, while looking through the telescope. The
adjustment of the eye-piece to the wires, and of the wires to the
focus of the object-glass, so as to avoid parallax, requires a nicety
which practice alone will impart to the observer; and which must
always be repeated till it is accomplished.

That the weight of the telescope may not cause the horizontal
axis to bend, the latter is made of two cones, B B, whose bases
are connected together (to form the axis) and to the telescope by
the intervention of a sphere, F, through which the telescope
appears to pass; it in reality however, forms the nucleus of the
instrument, to which the tubes A and A′ forming the telescope, as
well as the two cones forming the axis B B, are attached. The apex
of each cone is finished with a steel or bell-metal cylindrical pivot,
turned and ground upon the axis as true as possible. Much of the
excellence of the instrument depends upon this being correctly and
well done, for if they are not true, the telescope cannot possibly
describe a vertical plane. These pivots work in v-formed sockets,
or Y's, as they are technically called, which surmount the upright
arms C C, of the iron stand.

To test the horizontality of the axis, a spirit-level, H, is placed
striding across the instrument with its standards resting upon the
pivots; for it is clear that if the pivots of the axis are horizontal (all
other things being correct) the telescope when elevated or depressed,
or turned completely over and pointed in the opposite direction,
must continue in the same vertical plane; when we thus say "the
telescope," it must be understood to mean the line of collimation,
or central intersection of the cross wires of a properly adjusted
instrument One or two thin brass plates are generally supplied in

the box with the instrument, which may be attached, by a milled-headed screw, to the top of the arms of the stand. These are notched at their tops to receive a pin fixed in the end of the tube of the level H, which thus prevents the possibility of the level falling off the instrument when in use.

The cross wires in the telescope generally consist of a single thread of a spider's web, which, at the same time that they are extremely fine, are perfectly opaque, and are not found to be fringed with light along their edges, as is sometimes the case when other fine material is employed. The volume of light which passes through the telescope in the day-time shews up these cross wires, but in using the instrument at night to range a distant lamp, or if need be, underground, the wires would not be visible, and, therefore, the instrument would be useless under such circumstances; to obviate this, and also to enable the possessor of such an instrument to amuse himself at his leisure, (if he does nothing more useful), with observing transits of the heavenly bodies, one of the pivots is perforated through the cone B, and sphere F, and the light from a lamp or lantern adapted to the top of the stand passes through the said perforation, and falling on a diagonal reflector, situated in the sphere, is thence thrown upon the focus of the instrument, where the wires are fixed, near to the eye-piece. The reflector is also perforated to allow of the passage of the cone of rays in their convergence from the object-glass to the said focus, and thus distinct vision, both of the distant object to be viewed and of the cross wires is obtained by night as well as by day. The detached figure 1 shews the lantern in its position when used as above described.

The iron stand is fastened to the stone after it is placed in the required position, by means of the screws K K, which work into sockets previously let into the stone; the holes in the stand through which the screws pass are *drawn*, to allow of a small motion being given to the stand to perfect its adjustment in position, before the screws are made tight.

THE ADJUSTMENTS AND FIXING OF THE TRANSIT INSTRUMENT.— Long before it is necessary to commence working operations, the

E

direction of the centre-line of the intended tunnel must be known nearly, and should have been staked out.

The most suitable spot must be selected for the Observatory, so as to command a view of the whole length of the work, and its erection forthwith proceeded with; also, if possible, a permanent mark should be set up at a distance from each end of the tunnel precisely in the intended line, for future reference and the occasional adjustment of the transit; these marks should be placed where there would be but little chance of their being disturbed or removed. For this purpose at Blechingley Tunnel I caused to be erected two brick piers, about five feet high, at the distance of full two miles from each end of the tunnel. There is an advantage in having them at a great distance, as far as accuracy is concerned in the adjustment of the transit, but in thick weather they cannot be seen, and, therefore, other similar marks should also be set up at shorter distances. All these piers were painted black, with a white line, from two to six inches wide (according to their distances), to denote the precise line with which the centre of the tunnel was to be coincident throughout its whole length. A straight line from the centre of one of the above-described distant marks, to the centre of the other (in the opposite direction), must pass through the telescope of the transit instrument when set for use, so as to coincide with the line of collimation. How to fix the telescope in a position to answer these conditions will be presently described.

To examine and adjust the line of collimation.—When the instrument is placed on its stand, direct the telescope to some small, distant, well-defined object, (the more distant the better,) and bisect it with the central intersection of the cross wires; then lift the telescope very carefully out of its angular bearings, or Y's, and replace it with the axis reversed; point the telescope again to the same object, and if it be still bisected, the collimation adjustment is correct; if not, move the wires one half of the error, by turning the small screws which hold the diaphragm with the cross wires near the eye-end of the telescope, and the adjustment will be accomplished; but, as half the deviation may not have been correctly estimated in moving the wires, it becomes necessary to verify the adjustment by moving the telescope the other half, which is done by turning the screws c c, near the top of one of the arms of the iron stand. Having thus again bisected the object, reverse the axis as before, and if half the error has been correctly estimated the object will be bisected upon the telescope being directed to it; if not quite correct, the operation of reversing and correcting half the error, in the same manner, must be gone through again until by successive approximations, the object is found to be bisected in both positions of the axis. The adjustment will then be perfect.

To set the Axis of the Telescope truly Horizontal.—Set the level H upon the pivots of the axis, (as shewn in the engraving), and by turning a milled-headed screw, a, near the top of one of the arms of the stand, raise or depress that Y, and with it the pivot that rests on it, till the spirit-bubble stands central in its tube; now reverse the level, that is, turn it end for end, and if the bubble again becomes central, it is clear that the axis is horizontal; but if it should not become central when reversed, it is equally clear that the axis is not only *not* horizontal, but that the level is out of adjustment also. To effect this twofold adjustment, half the error must be corrected by raising or depressing the screw a on the stand, and the other half by turning the capstan screws b b, at one end of the level, which raises or depresses the spirit-bubble with respect to its points of support that rest on the telescope pivots. It will, however, be obvious that the axis of the instrument will have been correctly levelled by the first part of the process, namely, *raising or depressing the screw a*, even though the error of the level be left untouched; but it will be found convenient to correct the latter error at the same time, as above described. These corrections, like those for the adjustment of the line of collimation, frequently require to be made several times before the adjustments are satisfactory; and when perfect, the spirit-bubble will remain central in the glass tube, both before and after its reversion. A graduated ivory scale is fixed along the top of the level, by which the amount of deviation from horizontality can be more correctly determined, and which scale for astronomical purposes is made to denote the angular deviation of the axis at the moment of observation. The value, in arc, of each division having been previously determined, a correction due to such error may subsequently be computed, and applied to reduce the observation to what it would have been had the axis been truly horizontal: but such minutiæ do not enter into the business of a Mining Engineer.

To fix the Transit-stand on the stone pier, and set the Instrument for use.—The instrument being set on the stone, move it, stand and all, until the telescope very nearly coincides with the two distant marks when directed alternately to them. Having thus approximately placed it, set up the level H, and adjust for horizontality (the collimation adjustment having previously been verified); now move the instrument very quietly till the telescope coincides with one of the distant marks, keeping the axis horizontal throughout this part of the business; then turn the telescope over, and look in the other direction,—(the observer must alway remember to remove the level before he turns the telescope over, or he will most probably injure, if not destroy it);—see if the coincidence with the other mark is correct; if it is not, observe the amount of deviation, and as nearly as can be judged move the stand laterally (or sideways) to correct one half of the deviation, and, by gently pushing one end of the stand from you, correct the other half. If these movements have been made judiciously, the telescope will be found, upon reversion, to cut both the distant marks; or otherwise they must be repeated till it is accomplished *very nearly*, so as easily to be perfected before the stand is finally screwed down, which must next be done.

The position of the instrument having thus been approximately settled, mark on the stone the position of the screw-holes in the bottom of the stand, and a mason can let into the stone the screw-sockets, which may be fastened with melted lead, plaster, or cement. This done, again set the instrument up, and insert the screws K K into the sockets, through the holes in the iron stand; which again approximately places the instrument in position. This approximation being now made very close, by repeating the foregoing operation, the final touch may be put to it, by first carefully examining and correcting its horizontality ; making the cross wires intersect one of the distant marks most carefully, by turning the capstan-headed pushing-screws *c c*, on the top of the other arm of the stand. This gives a horizontal motion to the Y, and hence to the pivot which it carries : now reverse the telescope, by turning it over on its axis, and see if the other distant mark is also bisected ; if so, all is well; if not, correct half the error by the capstan screws *c c*, and gently *tap* or *slide* the stand laterally the other half;—the holes for the screws K K being drawn or made sufficiently large to admit of this small motion being given to the stand. This adjustment, like all the others, sometimes requires repeating, especially if not done with delicacy and care. When done, the screws K K may be made fast.

The method now described of bringing the instrument exactly into line may appear rude, and difficult to accomplish, as it requires a nicety of touch which experience in the use of mathematical instruments (almost) alone imparts to the observer. To overcome this objection, by imparting mechanically a lateral motion to the stand, some of them are constructed with a double frame : in which case, the lower one is first firmly fixed to the stone (approximately in position), and the upper one is then rectified by means of screw adjustments, not unlike one of the motions of a *slide rest* :—which is a great convenience, but adds considerably to the cost.

When the stone is being fixed on the top of the brick pier, it should be set correctly level on the upper surface; otherwise much trouble will result in subsequently fixing the instrument.

The foregoing are the adjustments necessary upon the setting up of the Transit Instrument; and, although it be ever so correctly done, it will require to be verified each time the instrument is used for ranging the lines, to ascertain that no derangement has taken place. To prevent this as much as possible, except from causes over which the observer has no control, he should lock up his observatory when he leaves it, and keep the key in his own possession, to prevent the ingress of ignorant and curious persons, who are too frequently prompted, although unintentionally, to do mischief.

A Transit Instrument is of all others the best adapted for ranging straight lines : and for such purposes, where great accuracy has been

required, they have been used for a long time passed: as in the
setting out of the base line upon which the great trigonometrical
surveys have been founded, both in this country and in distant parts
of the globe. Likewise in surveys of small districts. Even in the
preparation of parish plans, after the manner directed by the Tithe
Commissioners, an instrument of this kind is highly useful for ranging
accurately the long lines intended for measurement, for it is not
sufficient that the measurements be correct, if the lines have not
been set out perfectly straight.

In setting out about fourteen miles of the South-Eastern Railway,
from Red-hill in Surrey to Chiddingstone in Kent, through both a
hilly and thickly-wooded country, the author found the use of the
Transit invaluable for the purpose. The railway, through this portion,
consists of a few long straight lines which lie nearly in the same
straight line; and having found a few given points on elevated spots,
he was enabled, by the superior power of the Transit Instrument, to
range each straight line both ways for long distances, with the
greatest precision. The instrument used on that occasion was of
small magnitude compared with the one described in the preceding
pages: it was portable, and adapted to a stout tripod stand, which
ensured its steadiness; and when removed from the stand it could be
placed in a conveniently formed box, and carried from spot to spot
wherever its use was required. The top of the stand had also a
lateral motion, to enable the centre of the instrument to be brought
readily over any precise spot on the line. Theodolites are now sometimes made with a transit telescope; which gives to those valuable
instruments a great additional advantage.

The possessor of a Transit Instrument may likewise employ it for
astronomical purposes, by placing it on a small brick pier in a garden
or yard, and temporarily covering it over, to preserve it from the
weather. For details of its practical application to such purposes,
the reader is referred to a " Treatise on the Principal Mathematical
Instruments employed in Surveying, Levelling, and Astronomy," by
the author of this work.

CHAPTER III.

SETTING OUT THE SHAFTS.—RANGING THE LINE BOTH ABOVE AND UNDER GROUND.—TAKING THE LEVELS, AND ESTABLISHING BENCH-MARKS, OR POINTS OF REFERENCE.—ETC.

WHEN the Transit Instrument is in perfect order, as described in the last chapter, the centre of each shaft may be ranged, and staked out, so that they may be all in a perfectly straight line. The number of shafts, and hence their distances apart, will depend upon the speed with which the work is required to be executed. In ordinary kinds of strata, if the work is required to be completed in twelve months, the interval between the shafts should not exceed one hundred yards; but I believe that about two hundred yards has been the distance more generally adopted. It will be found convenient to have the shafts equidistant from each other, unless some peculiar form or circumstances of the surface call for a different arrangement.

As soon as the shafts are sunk to their proper depth, (the mode of doing which is described in the following chapters), it becomes necessary to transfer the ranging of the line from above to below the surface of the ground, either for making the large excavation of the tunnel, or for driving a smaller heading. When the ground is quite dry and sound, there is not always a necessity for the latter, but the excavation of the tunnel may be proceeded with at once. If, however, there be any quantity of water to contend with, as is too frequently the case, a heading at the level of the bottom of the intended work, to serve as a drain, becomes essential; and indeed, under all circumstances, a heading sufficiently large for a man in a stooping posture to pass along, will be found very convenient to ventilate the works, to form a ready communication from shaft to shaft, and to ensure accuracy in the levels and ranging, or setting out, of the work. In this way both objects were attained with the greatest certainty, at

Blechingley and Saltwood; for in no one of the junctions, could any deviation from accuracy be detected.

A common method of transferring the line from the surface to the underground works, is by the erection of a ranging-frame over each shaft; consisting of three half-timbers framed as a triangle, and supported at the angular points by stout props. In the engraving of the works above ground at Blechingley, (facing the title-page), a ranging-frame is shewn at the shaft on the right hand of the picture, and some men are represented on the top of the head-gearing of the gin, adjusting the lines according to the telegraphic signals from the observatory. The men stand quite free from the ranging frame, lest they should by moving it, displace the lines. One side of the frame (or triangle) is fixed parallel to the line of the tunnel, and on each of the other two sides is spiked a triangular prism-shaped piece, or block, of wood, in such a situation that a line passing over the projecting arris or angle of such block shall be in the intended line of the tunnel, and pass down the shaft as close to its side as possible, without touching the brickwork in its descent. The upper end of the line is fastened to a nail at the back of the timber, and a heavy plumb-bob is suspended below, to stretch it perpendicularly. The bob is immersed, or hung in a vessel of water, the more readily to bring it to rest, and ensure its steadiness.

The line that I used for this purpose was common fishing-line, rather larger than whip-cord. I am aware that copper wire, and other substitutes, have been tried, but believe they have not generally been found to answer better than the common line, which has the advantage of presenting less temptation to be stolen. That so fine a line may be distinctly seen through the Transit at a distance, a board, having one side painted white and the other black, was held up behind the suspended line, and when the white side was turned towards the observer, the line was rubbed over with charcoal, whereby the observer had to view a black line upon a white ground; and when the black side of the board was used, the line was rubbed over with chalk, when a white line upon a black ground was presented to view. Sometimes the one, and sometimes the other, was more easily to be seen :—for the shafts near the Observatory, the black line on the white ground generally answered best; but for greater distances the white line upon the black ground was preferred. This might probably in some measure thus arise. The wires in the telescope being black, could not so well be distinguished on the dark surface of the board when it was near, but this effect would diminish when the board was further off;—the state of the atmosphere might also have some influence on these different appearances. Two lines were necessarily suspended down each shaft, on its opposite sides; the one farthest from the Observatory was usually ranged first, and then the nearer one; after which they were both presented to the observer's view at once, by way of test, when, if the previous ranging of each line separately had been quite correct, both would appear as one line; and if such was not the case, a repetition of the process of ranging each line separately took place, until the result was satisfactory.

When the signal had been given that each line was correct, a notch was cut in the vertex of the triangular block where the line passed over it, which notch represented the correct point of suspension of the line, and formed a recess for it to rest in, retaining it in its proper place. A screw with a capstan head has been applied to this purpose, which cannot readily be disturbed, as persons have been mischievous enough to alter the notches, or cut more of them, to annoy the parties who might next use the line. These screws no doubt answer that purpose very well; but where the ranging-frame is erected on a spoil bank which is new-made ground, a continual settlement is taking place, and, consequently, a motion arises in the parts of the ranging-frame which would put the best contrivance out of adjustment: therefore, as it is necessary frequently to examine the notches over which the line passes, a wooden block will be found to answer all practical purposes. The annexed cut shews the appearance of the diagonal or cross wires of the telescope when bisecting the line suspended from a ranging-frame. The vertical line $a\,a$ is the ranging line, which must be moved to the right or left till it bisects both the upper and the lower angles of the cross wires, $b\,b$ and $c\,c$, of the telescope.

The foregoing method of ranging the lines cannot be satisfactorily used except in calm weather; for the wind has so great an effect in forcing the lines out of the perpendicular, even with a plumb-bob weighing nearly a quarter of a hundred weight suspended by it, that when the wind is at all high it is next to impossible to adjust them correctly, and therefore a delay from this cause will frequently take place in so uncertain a climate as ours, which is not only inconvenient, but, where the ground is heavy, may be attended with danger. I allude to a delay in setting the leading frames when the ground has been excavated ready for the brickwork. These circumstances led to my adoption, at Saltwood Tunnel, of the iron spikes with holes in them, figured and described in the next page.

When the heading or tunnel is to be commenced, it is necessary to mark, on opposite sides near the bottom of each shaft, the exact position of the intended centre line, that in advancing the works in those directions there may be a certainty not merely of *meeting* the workings in the opposite directions, from the other shafts, but that the meeting (or Thirl, as it is called,) may coincide correctly ; or, in other words, when the tunnel is completed, a line stretched from end to end should pass exactly along the centre of its breadth in all parts. Such a degree of accuracy is not always attained, neither is it so essential for the heading as it is for the tunnel ; it will be sufficient for the former if it be so straight that a line stretched along its whole length may be quite free (by a few inches) from touching its sides ; but the junction of the tunnel should be strictly correct, and may be so when proper attention is bestowed in ranging the work and setting the leading frames ; the latter is too often left to merely intelligent labourers, instead of being done by the Resident Engineer himself, and therefore, it is not surprising that we occasionally hear of such deviations from accuracy as six, twelve, and even eighteen inches in the meeting of the work at the junctions of various tunnels. To mark at the bottom of the shafts the direction that the workings are to take, is simply the transference of the position of the lines, when suspended from above, to the sides of the shafts, by ranging, and driving spikes or nails ; the lines having been previously adjusted to position, in the manner before described ; or by a more simple method, as follows.

In setting out the shafts for the construction of Saltwood Tunnel, I adopted a plan which obviated the delay arising from windy weather, &c., and saved much time, and chance of error afterwards: it was as follows. When ranging the intended centres of the shafts, (which were nine feet clear diameter,) a substantial stake was driven or fixed securely into the ground, about sixteen feet on each side of the centre of every shaft, and over the intended centre line of the Tunnel, so that when the shafts were sunk and bricked, these external stakes were about ten or eleven feet from their inner edge. On the top of each stake we drove a spike, made in the form shewn in the margin (one-half the real size,) and ranged the centre of the hole in the spike accurately with the Transit Instrument; this done, wooden caps were screwed over the said spikes, to prevent their being disturbed. The ranging of each spike can be done, with the greatest precision, by holding a piece of white paper at a little distance behind it, so that the hole may present a neat white disc for bisection with the cross wires of the transit; and, by thus watching the spike as it is being driven, any deviation to the right or to the left may be corrected by the man who is driving it, upon giving him the necessary signal with the telegraph.

A line is stretched centrally across the mouth of the shaft, and its ends passed through the holes in the spikes above described; it is then to be drawn tight, and made fast. A plank should next be fixed at each side of the shaft, at right angles to the line, and so placed that one side of the plank may hang over the shaft about two or three inches, or sufficiently to keep the line clear of the brickwork when it is suspended from the said plank. The lines may next be lowered and the plumb-bobs attached at the bottom, and when steady, the lines may be moved along the edge of the plank until they hang precisely under the horizontal line which crosses the shaft; in this position they may be secured by a nail driven into the plank, and the line twisted around it.

By the adoption of the ranging spikes, as above described, it is clear that no annoyance can arise from windy weather, which so much affects the lines when suspended from a ranging-frame elevated above the shaft,—because they are wholly sheltered; and, if care be taken to keep the ranging spikes from disturbance, the lines can be dropped down the shafts at any and at all times, night or day, for setting the leading frames, or verifying the position of fixed marks: an advantage too well known to the practical man to require further comment.

When the lines and the suspended plumb-bobs have come to rest, and are steady, nails may be ranged by them, and driven into the square timbers on each side at the bottom of the shafts; thus establishing two points in the line required, from each of which the workmen may suspend a plumb-line to guide them in keeping their work in line, and by which they may also correctly fix other nails more distant from the shafts, as their works recede therefrom.

Some attempts have been made to range the lines below by means of the Transit Instrument itself; for which purpose it was placed over the shaft, and having been correctly adjusted to position, &c. by some such means as have been previously described, the telescope was turned vertically downwards, whereby the observer, (by standing over the instrument,) might look down the shaft, (the base of the stand being made open for that purpose); and thus it was supposed he might range a line, a wire, lights, enamel discs, or any other of the many contrivances which have been suggested for that purpose. Such methods are pretty in theory, but in practice not very likely to answer; for independent of the difficulty of thus accomplishing the required object, much time would be consumed, and the result not so satisfactorily arrived at as by the method of dropping the lines from the ranging-frames; and less advantageous, both as to expedition and certainty, than the method by the ranging spikes adopted at Saltwood, as before described.

As the miners advance their headings, or recede from the shafts, they must fix (as before stated) other ranging points from which to suspend more plumb-lines; this they do, by looking back, and ranging them with the lines in the shaft; and from these new points they range others still more remote, to prevent their deviating from a straight line. When the works at Blechingley had advanced to this stage, the continual verification of the men's proceedings became

necessary; and for this purpose the author contrived a candle-
holder, (shewn in the annexed engraving,) which
answered extremely well. It was previously
the custom when ranging several lines at short
distances apart, for a man to stand near each of
them with a candle, which he shaded with one
hand to keep the direct light from the observer,
and to throw more light upon the line: this
occupied the time of several men, and was not
so satisfactory as the use of the candle-holders
which were suspended from the same nails that
the miners attached their lines to. Four of
these were generally used at one time, and by

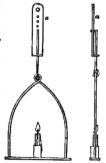

raising or lowering them in their racks, a, the flames of all the
candles could be brought to the same level; and if the nails from
which they were suspended were in the same straight line, all the
four candles would appear but as one, when viewed in a proper
direction; and, on the other hand, the least deviation from the
proper line in either of the nails under trial was distinctly shewn,
especially in the otherwise darkness of the underground works; and
therefore it could be corrected. This statement may easily be
proved, by observing how correctly any number of candles, of the
same height and on the same level, can be ranged in a line along a
passage or corridor, or even on a table. By such means the work of
the miners was kept straight. The heading at Blechingley was
remarkably correct: that at Saltwood was not so well done, arising
from the difficulties to be contended with, as
already mentioned, and will be hereafter more
fully described; nevertheless, a line could be,
and was stretched through the work without
touching the sides, being nowhere within nine
inches thereof, and therefore it answered all the
purposes required.

The annexed cut shews one of the candle-
holders suspended from a nail in the upper
cross-piece or cap, B, of the heading frame, or

" setting," as it is termed. The upper part, or rack, a,—(see last figure)—was made of thin sheet iron, with a number of holes in it; the remainder was of iron wire, carrying a socket for the candle. By means of the rack, the candle could be raised or lowered to the proper level, and being hung by a flat plate it was prevented from revolving and thereby interposing the wire in the line between the flame and the observer, which would frustrate the desired object. This position is shewn in the right-hand figure of the former engraving.

As soon as the headings were driven from end to end, the permanent ranges were fixed; each consisting of a cross-piece, a,— (see last figure)—fixed to a setting in the heading, at intervals of about thirty or forty feet, and having marked thereon where the intended centre-line of the tunnel would cross, a block of wood, b, was screwed down, having a hole through it which was placed in coincidence with the said centre-line mark. By passing a line through these holes in succession, the centre-line of the tunnel was ranged at all times. The method of determining the position of the centre marks on the cross-pieces is by suspending the vertical lines, as before described, down two or more consecutive shafts, and stretching a long line very tight in the heading; which line is then moved to the right or left, until it coincides with or is perpendicularly under the two lines in each shaft at the same time, and where the line then crosses the piece of wood, a, is the position for the central mark, and the hole in the block, b. When those marks that are near to the shafts are removed by the construction of the side and first leading lengths of the tunnel, it is convenient to fix in the invert a post of balk timber, with an iron cap, having a similar hole to those in the blocks above named, centrally arranged, through which to pass the line when required. When these marks are fixed all the way through the heading, a number of points will be established in the centre-line of the tunnel, and therefore the whole, or any portion of the work can be correctly ranged out whenever required. With such conveniencies at hand, no leading frame should ever be set but by this line to denote the position of its centre. No further reference need be required to the surface, or to the Transit Instrument; which, however, had better be kept up until near the completion of the work, or until all chance of its being again required has passed away, the Observatory may then be taken down, and the materials worked up. When, however, the tunnel is to be driven without the use of a heading, frequent reference to the points on the surface will be required.

In perusing the above particulars, it will strike the reader, that if the line stretched along the heading from whence the central holes in the blocks, *b*, are derived, extended the whole length of the tunnel at once, the greatest accuracy in the result would be obtained: but such a line, when the tunnel is long, would 'sag' too much, and probably break before it could be strained sufficiently tight; therefore a line as long as possible, so as to answer these conditions, should be used, and thus the whole length be ranged piecemeal; which, having been done with care, and embracing in each length at least two of the previously ranged centres, little or no error need arise.

The whole of the foregoing methods and details of ranging the lines were not adopted by the author at the first starting of the work; but suggested themselves from time to time, as circumstances arose. They are given as the methods he most approves, and would adopt in future, if ever again called upon to execute similar works.

THE LEVELS AND BENCH-MARKS.

Having described the method of keeping the work straight in a horizontal direction, it remains to explain that of making it correct in a vertical direction; or, in other words, preserving the proper level. When the tunnel is upon the same level throughout, the task is easier than when it is inclined; although the latter presents no difficulty, or particular chance of error, to a careful person. A section of the ground must first be made along the intended line of the tunnel, and the relative level thereto of some standard bench-mark determined for future reference. The position of the shafts having been determined upon, (or even began), a substantial bench-mark should also be established opposite to each shaft, but at such a distance therefrom as to be a little beyond the area which the spoil-bank and materials used in the course of the work are likely to cover, so that it may be always accessible for reference. For this purpose, large stakes or ends of square timber should be driven or fixed in the ground, so as not to be easily disturbed, and an iron spike with a round head driven in the top, upon which the levelling staff can be held; and they may be denoted by the same numbers—1, 2, 3, &c.—as the shafts to which they are opposite. The relative level of these bench-marks with respect to that of the Railway, at a point immediately under their opposite shafts, must be determined, and hence, by computation, its height above some given point in the tunnel may be found,—(the springing, or skewback of the invert was the point to which preference was given for this purpose)—and registered for after reference. Thus, at No. 3 shaft at Blechingley, the height of its bench-mark above the standard datum of the Railway section, as found by levelling, was 343.78 feet; the formation level of the Railway at a point immediately under the said shaft, as determined by the section, and computed on the intended gradient, was 249.99 feet above the said datum: therefore 343.78—249.99=93.79 feet for the height of No. 3 bench-mark above the formation level under the centre of No. 3 shaft. But the formation level was 1.25 below the intended level of the skewback; therefore 93.79—1.25=92.54 feet for the height of the bench-mark above the intended skewback of the tunnel. This method of computation was adopted for every shaft, and registered for future use, as in the following table :—

Shaft.	Formation Level above datum at Shaft.	Bench-Mark above datum.	Bench-mark above Formation.	Formation below Skewback.	Bench-Mark above Skewback.	Staple in Shaft.	
						Above or below Bench-Mark.	Above Skewback.
	feet.	feet.	feet.	feet.	feet.	feet.	feet.
1	250.37	320.87	70.50	1.25	69.25	— 2.23	67.02
2	250.18	330.38	80.20	.	78.95	+ 0.22	79.17
3	249.99	343.78	93.79	.	92.54	+ 2.70	95.24
4	249.81	339.64	89.83	.	88.58	+ 7.39	95.97
5	249.67	330.68	81.01	.	79.76	+ 8.61	88.37
6	249.52	334.94	85.42	.	84.17	+ 2.73	86.90
7	249.37	335.28	85.91	.	84.66	+ 5.71	90.37
8	249.21	328.85	79.64	.	78.39	+ 3.63	82.02
9	249.05	322.98	73.93	.	72.68	+ 7.35	80.03
10	248.89	317.54	68.65	.	67.40	+ 7.09	74.49
11	248.72	302.33	53.61	.	52.36	+14.71	67.07

All the dimensions connected with the sinking of the shafts may be taken sufficiently correct with the common measuring tape; but for the actual tunnelling operations, more certain means must be employed. It is advisable, in the first instance, to fix a bench-mark at the bottom of every shaft, which shall represent the intended level of the skewback of the invert, or any other level that may be chosen. A flat iron spike driven into the timber at the bottom of the shaft, and projecting sufficiently for the level staff to be held upon, will answer very well. Having drove this at every shaft, the Engineer should level through the heading (if there is one) from shaft to shaft, throughout the whole extent of his work. By this means he will prove how far he is accurate, and if as he proceeds along the heading, he fixes at short intervals similar bench-marks at the proper level for those points where they are fixed, he will have the means of subsequently checking his work as it proceeds; and will relieve himself of all doubt as to the final result of the levels at the several junctions when they may be effected.

The method adopted for transferring the levels at the surface to the bottom of the shaft, (or dropping the levels, as it is usually called), is to drive securely into the brickwork on the inside and near the top of the shaft a stout horseshoe-shaped staple, the circular part of which is left projecting from the brickwork, and forms a loop for a wooden rod to be passed through to the bottom of the shaft. The height of the upper edge of the staple above the intended works below, having first been determined by means of the neighbouring bench-mark, as shewn by the two last columns of the above table, is the point from which the required length of rod is to be suspended, so that the bottom of the rod may represent the level of the invert skewback, or whatever other level may have been fixed upon. An iron gland is attached to the rod, at the proper point, which gland is larger than the loop-hole of the staple, and therefore rests upon it, and suspends the rod as required.

The wooden rod above spoken of must, from the great length required, consist of a number of pieces which are attached to each other by various contrivances; some of them have been made to screw together, precisely as the old-fashioned round legs of a spirit level were joined in their middle; but, having previously witnessed considerable inconvenience and loss of time by this method, I caused my rods to be connected with a spring catch, or hook and eye, nearly similar to the hooks we used for suspending the skips from the windlass ropes for the purpose of lowering them down the pits.

The annexed engraving shows the rods used upon the works, their connecting hook, and the gland. The rods are shewn by the left-hand figure, with the gland attached at the point B. The upper right-hand figure shews the hook, the corresponding eye, and the ends of the rods to one-half the real size. The lower right-hand figure shows the form and make of the gland; A A are the cheeks which clasp the rod; C is the screw for tightening or releasing them, which screw is worked by a handle, D, similar to that of a vice; B is the hinge by which the cheeks open to clasp the rods, and when open the screw C is drawn quite out of the socket.

The rods were made ten feet long from the inner edge of the loop or eye, at one end, to the inner edge of the hook of the other, or from the two points of connection; and when they were suspended in the shaft, each ten-feet rod became as it were the link of a chain, consequently the whole would be sure to hang perpendicularly, which was not often the case with those that screwed together. Each rod was numbered, for convenience in use, and was hooked on the top of the one previously lowered; number 1 having been first lowered, the succeeding numbers showed how many tens of feet had been passed down through the staple; thus, when number 6 was hooked on, it was evident that 50 feet of rod were already down; and when any odd number of feet and decimals of feet had to be suspended from the staple, as for instance 95.24, the nine rods were first lowered, and upon the tenth the odd 5.24 feet was marked, and at that point the gland was screwed to clasp the rod firmly, the underside of the gland coinciding with the said mark; the tenth rod was then hooked on to the ninth, and likewise passed down until stopped by the gland resting on the staple, and the whole then hung perpendicularly.

In all cases, both at Blechingley and Saltwood Tunnels, when the rods were suspended in the shafts, the bottom of No. 1 rod represented the required level of the invert skewback (or springing of the inverted arch), which was the standard there adopted, and for the greater convenience of transferring the said level to any other point, as a ground-mould or bench-mark purposely made, the lower three feet were graduated exactly as the levelling staff was divided; the rod was then illuminated with a candle wherever the line of sight of the level cut it.

The method of transferring levels from one point to another being so simple an operation, the reader is doubtless well acquainted therewith; and therefore no further explanation of that matter is required. In all instances the rods were suspended at least three times down every shaft; first, for establishing a number of bench-marks through the headings, as before stated; secondly, for setting the ground-moulds for every side length; and thirdly, when the shaft lengths were completed, to fix an iron bench-mark in the brickwork exactly at the proper skewback level, at the side of the tunnel under each shaft; for although the skewback at that place had then been built and completed, yet it was considered necessary to fix a proper point on which to hold the levelling staff at the exact calculated level, for the brickwork (however well done) would be almost sure to vary a small quantity therefrom. These bench-marks consisted of flat iron spikes, projecting about an inch from the wall into which they were driven; and from these standard skewback levels all the subsequent ground-moulds were set, making the calculated allowance for the rise or fall of the gradient; which would be unnecessary where the tunnel is to be constructed on the same level throughout.

.

CHAPTER IV.

SHAFT SINKING.

THE TRIAL SHAFTS.

THE construction of the Blechingley Tunnel commenced with the sinking of two trial shafts, to ascertain the character of the strata through which the tunnel was to pass. Instructions for their commencement were given on the first of February, 1840, and on the third a contract was made, with two men, to sink and brick the shafts in question. They were to be six feet diameter in the clear; the brickwork to be nine inches in thickness; the bricks to be laid all headers, properly breaking joint, and to be set in the best greystone lime mortar: the price was to be,—for the sinking, eighteen shillings per yard down, and for the brickwork at the rate of sixteen pounds per rod; which prices were founded upon the following estimate.

	£ s. d.	£ s. d.
EXCAVATION:—		
4.9 cubic yards in 1 yard down ... at 2s. 6d.	0 12 3	
4 oak curbs for a depth of 33 yards, with plates and bolts complete ... £2 each, or per yard down	0 4 10	
Props, chogs, spikes, nails, candles, &c. ...	0 1 0	
Excavation per yard down ...	... ...	0 18 1
BRICKWORK, per rod:—		
Bricks ... 4,500 ... £2 2 0 per thousand		
Carting ditto ... 0 10 0 ,,	11 14 0	
2 12 0		
Lime .. 1¼ yard ... at 13s. ...	0 16 3	
Carried over ...	12 10 3	0 18 1

	£	s.	d.	£	s.	d.
Brought over ...	12	10	3	0	18	1
Sand ... 3 yards ... at 2s. 6d. ...	0	7	6			
Labour, candles, &c. ... per cubic yard, 5s. ...	2	16	7			
Per rod ... =	15	14	4			
and ∴ per cubic yard =	1	7	9			
and 1.8 cubic yards in 1 yard down ... =				2	9	11
Total Estimate per yard down ... =				£3	8	0

The situation of the trial shafts was the same as that of the working shafts Nos. 1 and 10—fig. 1 plate 1—as they were subsequently enlarged to nine feet diameter to make working shafts of them. They were designated as the western and eastern trial shafts; No. 1 being the western, and the other which afterwards became No. 10, the eastern. They were both commenced on the seventh of February, 1840, and on the tenth the western shaft was sunk 14 feet, and the eastern 8 feet. My memorandum on that day was as follows:—

" At present, the earth shews no indication of water below the first six feet; it is a kind of marl, of a slippery or greasy feel, something resembling fullers' earth."

On the twelfth, the western shaft was sunk 25 feet, when it was no longer safe, or scarcely practicable, to proceed without inserting the brickwork, as the last six feet had yielded a considerable quantity of water. Moreover, by exposure to the atmosphere, particularly if any damp was present, the ground slaked, or rather dissolved, accompanied with a swelling or heaving movement. This was invariably the case throughout the subsequent operations, and therefore it was unsafe to leave a face long exposed to atmospheric action, as such heaving brought a great weight upon the work, and in few cases, where the excavated lengths remained longer than usual for the brickwork, the weight became such as to break the bars; which, on this account, were provided of large dimensions, averaging 14 or 15 inches diameter for the crown, and 12 inches for the side bars, all of oak. The brickwork was commenced upon an oak curb fairly bedded on the bottom of the excavation, and set level; the inner diameter of the curb (or ring) was that of the intended shaft,—6 feet; its width, 9 inches, to carry that thickness of brickwork; and its depth, or thickness, 3 inches: it consisted of several pieces, connected by half-lap joints, having an iron plate crossing the joint on each side of the curb, and secured by four bolts, which passed through the timber and both of the iron plates. Upon this curb the brickwork was carried up to the surface, being well packed and rammed with dry earth, wherever there was a vacancy at the back, to make it solid.

By the time the brickwork was finished, about 14 feet of water had accumulated in the shaft, which had to be drawn out before the work could be resumed. When the sinking had advanced two feet further, a bed of hard calcareous sandstone, about one foot in thickness was met with;—this required blasting;—and it was separated from a still harder rock below by about one inch of sand; this lower rock proved to be two feet in thickness; next followed eight inches of brown sand, and three inches of bluish-grey sand; after which the clay or shale, the same as that we passed through above the stone, reappeared. It was expected that the stone formed a regular bed in the above situation, where it would have given considerable trouble as it was nearly level with the top of the tunnel; it however occasioned only a temporary difficulty, as it soon disappeared, probably from some dislocation of the strata, as the hill was full of faults, and had undergone great derangement. This was evident from the strata lying most abruptly in all directions, as was strikingly shewn in the open cutting at the west end of the tunnel, and mentioned at page 7. It is probable that the rock was an outlier from the lower bed of the lower Green Sand; the Weald Clay in which we were working cropped-out from beneath that stratum, about half-a-mile to the north.

The water henceforth flowed into the shaft so fast that but little progress was made with the sinking; and during the week ending February 22nd no more than 7 feet 6 inches was excavated; another oak curb was then placed, of the same dimensions as the former one, and vertically under it; the upper one, with its load of brickwork, being supported by raking props, which were removed as soon as the brickwork was carried up from the lower curb to the under side of the upper one, which was thus under-pinned, and the whole formed one shaft. This second length of brickwork was set in cement, on account of the water. Whenever the regular courses of brickwork brought up from the lower curb would not exactly fill the space, or fit tight to the underside of the upper curb, it was generally done by one course of bricks set on edge; and any small spaces that might then be left were filled up by driving in oak wedges.

With considerable effort a third length of brickwork was got in, and similarly secured, and a fourth length of excavation was commenced. On February 29th I reported as follows:—

" By perseverance, and at a considerable expense in drawing water day and night, together with shoring and poling the earth to prevent its falling in, I have succeeded in sinking the western shaft to the depth of 40 feet 6 inches ; at this depth so great an influx of water came upon us, that we scarcely had time to prop the brickwork before we were compelled to leave the shaft ; this water has continued to flow so fast that after obtaining larger buckets, and six men drawing day and night, it could not be kept under ; and at length, finding that it gained upon us I abandoned it as fruitless without more powerful means. I have therefore left the shaft, until by driving the heading we shall drain the springs : the water is now within two feet of the surface ; there being full 38 feet of water in the shaft."

Before finally leaving the shaft, sufficient earth was thrown in to fill up the excavated part below the brickwork, and a foot or two higher, in order to prevent the earth coming down from the back of the brickwork, which would have endangered the stability of the whole shaft.

The shaft thus left, was soon filled with water, which ran over in a considerable stream. It was then fenced in to prevent accidents, and remained in that state for several months, as a dispute with the occupiers of the land about their compensation, delayed our getting possession thereof until the following August. When possession of the land was obtained, and we were sinking No. 2 working shaft, at a distance of 110 yards to the eastward of the above trial shaft, we were similarly impeded with water, and, after drawing it for a day or two, it was observed that the water in the trial shaft had ceased to run over, and had sunk a little; taking this hint, men were set to draw as hard as possible at both the shafts, and in a few days they were drained to the bottom; and, although we were afterwards more or less troubled with water, yet by drawing at the two shafts we could always prevent its stopping our progress.

The soil through which the eastern trial shaft (afterwards No. 10.) was sunk, was more nearly uniform, and consisted of clay or shale highly indurated; the first 25 feet were of a brown colour, the rest to the bottom was of a deep blue, with an occasional bed of stone about two inches in thickness. (The brown-coloured clay at the western trial shaft terminated at a depth of 18 feet.) We reached 48 feet before we came to water, and were able to sink to 59 feet (by drawing water), before we placed the curb and commenced bricking up. This was on February 26th, and the brickwork was level with the surface of the ground on the 5th of March; an interval of eight working days inclusive. During this period, seven feet of water collected in the shaft, which being drawn, the sinking was continued to the then full intended depth of 66 feet, which was completed on the 14th of March.

From what has now been stated, it will be evident that the work could not be done at the price at which it was undertaken; for in the estimated expense no such contingency as the occurrence of water was provided for; and furthermore, the carting of the bricks to the ground cost 20s. per thousand instead of 10s. as estimated; for none could be obtained nearer than seven miles, and by a hilly road. It justly fell upon the Company to pay these extra expenses, and the following will shew the actual cost of these shafts:—

	£ s. d.	£ s. d.
Total outlay, 35¼ yards done	202 17 2	
Deduct materials left for future work ...	20 15 3	
Total absolute Cost	... ...	182 1 11
Estimated cost of 35¼ yards	120 14 1	
Extra cost of carting 24,000 bricks, at 10s. ...	12 0 0	
		132 14 0
Extra cost upon the lower 15 ft. 6 in. of the Western Shaft, in consequence of water		£49 7 11

The men having taken to the work at a price per yard down, or running yard, reckoned at the rate of 2s. 6d. per cubic yard, it may be useful to shew the amount of their earnings, thereby to judge how far the price was a fair one. The greater depth of the eastern shaft, where no difficulty occurred that was not anticipated, will give a fair average; for it must be remembered that great progress can be made at first, which necessarily diminishes as the shaft gets lower. On February the tenth it was down 8 feet, and on the twenty-sixth, 59 feet,—a difference of 17 yards in fourteen working days, averaging 1.2 yards per day,—which, at 12s. 3d. per yard, amounted to 14s. 8d. per day, to be divided among four men, and if we consider the odd 8d. to be the cost of the candles, it leaves 3s. 6d. per day for each man, supposing their earnings to be equally divided.

From the sinking of the trial shafts it became evident that the stratum in which the tunnel was to be formed was the Weald Clay of geologists; it was also near its outcrop from beneath the lower Green Sand; it having been previously a question whether or not the spur of Tilburstow Hill, through which the tunnel was to be formed, was not included in the green sand stratum. The shafts further indicated that a considerable quantity of water would, in all probability be met with in the course of the work, and that the ground would be heavy; upon this information, the form most suitable for the tunnel and the proper dimensions for the brickwork, &c. had to be determined.

TRIAL SHAFT AT SALTWOOD TUNNEL.

The first operation after setting out the centre line of the tunnel, was to sink a trial shaft, to ascertain the character of the strata beneath the surface; this was commenced on the 25th April, 1842, and was situated 13 yards from the centre line of the tunnel, on the south side. It was originally intended, that it should not only be a trial shaft, but was to have been sunk (if possible) so deep as to answer the purposes of a well, to supply the works with water during their progress, in the event of a deficiency of that article.

H

The shaft or well was six feet diameter, clear of the brickwork, which was nine inches thick, the bricks being laid all headers. The sinking was attempted by means of a barrel (or drum) curb, which upon being undermined descended by its own weight and that of the brickwork (which was constructed upon the curb, and was carried up as the curb descended). In this manner a depth of 46 feet was attained, when the drum getting a little out of the perpendicular, it stuck fast and could be got no lower, and after a fruitless attempt to liberate it, was left in its place, and the sinking resumed beneath in the ordinary mode to a further depth of 20 feet 6 inches; when coming upon a quicksand, we were unable to prop up the brickwork during the process of underpinning, and therefore could proceed no farther by that method. Another curb was then constructed of the same diameter as the inside of the shaft, namely six feet, to continue the sinking from within the completed brickwork; but so large a quantity of water was given out, that eight feet additional depth only were attained, making a total of 75 feet down.

COST OF TRIAL SHAFT AT SALTWOOD TUNNEL.

	£ s. d.	£ s. d.
CARPENTRY :—		
Large drum curb	5 5 0	
Smaller ditto	4 10 0	
Platform for workmen in the shaft ...	1 1 0	
Eight curbs or rims ... 12s. ...	4 16 0	
		15 12 0
BRICKS : ... 17,000 ... at 51s. ...		43 7 0
Labour in excavating and bricking (or steining) 15s. per yard down		18 15 0
TOTAL		£77 14 0

The drum curbs, and the mode of using them, will be understood by reference to the annexed section of part of the shaft; which shows the first curb A in the position where it became immovable; the termination of the larger portion of the shaft at B; and the smaller curb C at the bottom of the shaft. The curbs were made smooth on the outside, that they might slide down easily;

were strongly bolted together, and the bottom edge formed like a wedge (or knife edge) to avoid resistance. Two plumb-bobs were suspended at right angles to each other, in the curbs, to guide the workmen in keeping them perpendicular. As the brickwork descended with the curbs it was continued upwards to the surface of the ground; thus constantly increasing the load upon the curb, and of course its tendency downwards, as the earth was removed from beneath.

There are various opinions as to the advantages derived from the use of barrel or drum curbs (running curbs, as they are sometimes called) over the ordinary mode of propping and underpinning. The latter mode is more generally adoped where the ground is sufficiently solid to carry the props in safety; in all other cases the drum curb may be advantageously employed; but it requires much care to prevent it from setting fast, as it did at Saltwood, which may principally arise from its getting out of the perpendicular, either from carelessness of the excavators, or from the earth yielding on one side; neither does it appear that the work can be more expeditiously done by its use. On the other hand, there is apparently a greater degree of security to the workmen, who being employed within the drum may perhaps be somewhat less exposed to danger; but this is even doubtful. To those who are unacquainted with the subject, the following particulars of the method of shaft sinking by props and underpinning may not be uninteresting:—

The shaft is first sunk to the full diameter of the outside of the intended brickwork, and as far down as the earth continues to stand safely. When it is no longer prudent to proceed, a timber curb or flat ring is laid upon the bottom of the excavation; this curb or ring has its inner or clear diameter the same as the intended shaft,

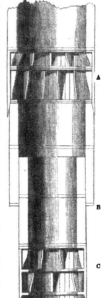

and its external diameter as much greater as twice the intended thickness of the brickwork, for upon its flat surface the bricks are to be laid. The curb should be made of durable timber, as oak, and formed in several segments, joined with half-lap joints, secured with a plate on each side, and four bolts passing through the whole; upon this curb the brickwork is carried up to the surface. The sinking is then renewed by excavating the earth from the centre of the shaft, as far down as may be consistent with safety, leaving a benching of

earth to carry the shaft, as shewn in the annexed engraving; a narrow portion of this benching is next cut away as far back as the brickwork, and a prop inserted raking either to the front or back of the intended brickwork; if to the front, the prop can be saved and used again, but it is sometimes necessary to place them raking behind the brickwork, in which case

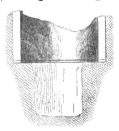

they are built in and lost. Another prop is then similarly inserted, and so on until the whole curb and brickwork, is thus supported. When this is done another similar curb is inserted perpendicularly under the upper one, and the brickwork carried up to meet or under-pin it. The work during this operation presents the appearance shewn in the following sketch. The props, however, are all shewn as raking inwards: one advantage in their raking outwards is,

that they leave more room for the brick-layers to work. The props may also be set perpendicularly under the upper curb, and the brickwork completed between any two props before they are removed; or the work may be quartered in, (that is, a quarter completed at a time.) Other methods of proceeding are also occasion-ally resorted to. The props must rest upon

a broad base, or foot-blocks, and be securely chocked to the curb above, to prevent motion taking place. As the brickwork proceeds, all vacuities behind should be rammed solid with dry earth.

In sinking through, and constructing shafts in the shingle beach upon the sea coast, at high-water mark, and also about midway between high and low-water marks, the following method of proceeding was found to answer. Having cast out the shingle and sand to a depth of five or six feet, or as far as conveniently practicable, a stout timber curb was laid level upon the bottom of the excavation, and upon this was built about four feet of brickwork, in cement; then, during the intervals of the tides, the shingle was removed from under the curb, by workmen within the shaft, and as they so removed it, the shaft gradually descended by its own weight, and the bricklayers continued to build it upwards, so as always to keep it above the level of the beach around; which otherwise would have filled the shaft at each high tide, and have occasioned great loss of time in its removal. This operation is similar to the sinking by means of a barrel curb, before described; but which would not answer so well in such loose material as that on the sea-shore, as it is capable of doing in firm soil.

On another occasion, the author was required to obtain fresh water for the supply of a large building erected on the sea-shore, to be used as an hotel. The spot where the building was erected had been left by the sea not many years before; the recession of the water,—or rather, the accumulation of shingle,—having been occasioned by the construction of a pier, for commercial purposes, extending into the sea close by. Upon an examination of the locality it appeared clear that within seven or eight yards from the surface, the top of the middle bed of the Lower Green Sand stratum would be found in situ; and therefore there was a reasonable probability that a supply of fresh water could be obtained from that level. But whenever a hole was made in the ground, which consisted of the beach and sand originally deposited by the sea, the salt water appeared and disappeared with the rising and falling of the tide, the sea percolating through the beach (which was there seventeen feet deep,) and rising in the hole to the level of the tidal water. Under these circumstances it was necessary to sink a shaft that should be water-tight, effectually to exclude the sea water from entering, and to prevent the fresh water that might rise in it from escaping; accordingly, a large shaft was first sunk, protected around with timber, by having square frames or settings at intervals of about three feet, and behind these, or between them and the earth (or beach), upright planks were driven close to each other, in the manner of sheet piling, some men driving at top while others were removing the earth from under the piles; this method was followed quite through the shingle, and answered well, but it could only be proceeded with at the time of low tide, as otherwise the salt water filled the timbered shaft. In a manner very similar to this, the difficulties in shaft sinking at Saltwood Tunnel were conquered; as will be explained in a subsequent chapter.

The timbered shaft was sunk quite through the beach and silt, and at eighteen or nineteen feet down, the middle bed of the Lower Green Sand was reached; when, as was expected, a large supply of fresh water was found, precisely in the same geological level as the water was met with in sinking the shafts at Saltwood. The brick shaft or well was then commenced upon a timber curb, sunk into the stratum about three feet below the bottom of the shingle, and nine-inch brickwork in cement carried up to the surface, and between the outside of the brickwork and the timber of the shaft the space was rammed with well-puddled clay, from the bottom to the top; the planks being left in the ground, lest their removal should disturb the puddling, and endanger the letting in of the salt water. The fresh water was admitted into the well through three pipes, which were built in the brickwork near the bottom of the shaft; but a short time afterwards, the pressure of an extraordinary high tide enabled the sea water to reach these pipes, and thus make the well-water brackish; whereupon the pipes were closed up, and the whole of the water from that level excluded, and by means of boring at the bottom of the shaft, a plentiful supply of pure water was obtained, which rose in the well to within four feet of the surface of the ground.

CHAPTER V.

SHAFT SINKING, CONTINUED.

EXCAVATING AND CONSTRUCTING THE WORKING SHAFTS, AND SUPPORTING
THE BRICKWORK BY SHAFT-SILLS AND HANGING RODS.

THE working shafts were 9 feet clear diameter, the brickwork
9 inches in thickness; an oak or elm curb was inserted at the bottom
of every length of brickwork as it progressed downwards, and at the
bottom of the brickwork, where the square timbering of the shaft
commenced, a curb of 4 inches in thickness was used; the upper
ones having been but 3, or $3\frac{1}{2}$ inches. These were to remain permanently in their places. As the sinking proceeded, notes were taken
of every change of strata, which, at the same time that they were
interesting as geological facts, were occasionally useful afterwards.
For instance, in sinking the trial shafts at Saltwood, we passed
through a stratum of clean sharp sand, well adapted for gauging
with cement. In the course of our subsequent work such sand
became a desideratum, and all we had to do was to break through
the brickwork of each shaft, at the particular level pointed out by
the memoranda, and a man threw down into the tunnel as much as
was required, excavating it as a driftway. Thus a plentiful supply
of excellent sand was obtained at a cheap rate.

Plate 2 shews two sections of a shaft at Saltwood Tunnel, fig. 1
at right angles to the tunnel, and fig. 2 in the direction thereof; the
plan adopted for suspending the brickwork, and square timbering the
lower part of the shaft is therein represented; likewise, the brickwork and the timber rings, or curbs, at a a, &c., at the bottom of
each length. The windlass, together with the skips for raising the
earth and lowering materials—the one ascending as the other
descends—are also shewn. The skips were suspended from the rope

by a particular form of hook, that prevented their unshipping in case of striking against each other in passing, or against the sides of the shaft. A further description of the above will be given in a subsequent chapter.

The following is a copy of the specification drawn up for the guidance of the bricklayers in constructing the shafts at Saltwood Tunnel:—

The brickwork of the shafts is to be nine-inch work, laid all headers. The shafts are to be nine feet clear diameter, and to be constructed truly cylindrical and perpendicular. The whole to be laid in mortar composed of one part of stone lime, and $2\frac{1}{4}$ parts of clean sharp sand, well mixed up with a proper quantity of water; the lime to be sifted before it is made into mortar. The mortar-joints to be sufficiently thin to make four courses of work not exceed one foot. The back of the work to be rammed and well punned, course by course, as the work proceeds, so that it shall not be possible to drive the bricks back from their places by applying any moderate force. Each brick to be dipped in water previous to its being laid. The bricklayer to find all tools, sieves, tubs, &c.; the engineer to find bricks, lime, sand, and water.

COST OF THE WORKING SHAFTS PER YARD DOWN.

	At Blechingley, in the Weald Clay.			At Saltwood, in the Lower Green Sand.		
	£	s.	d.	£	s.	d.
EXCAVATION: Including all tools, gunpowder, candles, and contingencies	1	10	0	1	2	0
BRICKWORK: Including all labour, candles, lowering materials, and contingencies	0	18	9	0	14	0
Bricks ... 1,020	3	2	0	2	11	0
Lime ... $\frac{1}{4}$ of a yard	0	4	3	0	4	3
Sand ... $\frac{3}{8}$ of a yard	0	1	4	0	0	0
	£5	16	4	£4	11	3

In addition to the above, the cost of the ring curbs must be added, which at both places was the same, and averaged about one in every three yards down.

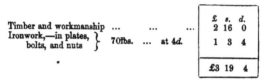

	£ s. d.
Timber and workmanship	2 16 0
Ironwork,—in plates, bolts, and nuts } 70℔s. ... at 4d.	1 3 4
	£3 19 4

The above details comprise the cost of the materials, and the labour actually consumed in the work; besides which, there were the windlasses, ropes, hooks, skips, planks, and props, &c., which were part of the general plant of the tunnel works, but a very small portion of their cost ought justly to be charged to the shaft sinking —they are included in the total cost of the tunnels.

SUPPORTING OR SUSPENDING THE SHAFTS.

The brickwork of each shaft was carried down to within a few feet of the intended top of the tunnel; and from thence through the space intended for the tunnel, the shaft was continued by means of timber only, having square frames or settings, at short intervals of depth, and close planking against the sides of the excavation; the square settings being supported from each other with props of round timber, (all of which are shewn in the two sections, plate 2, and in fig. 1, plate 3). This timbering was adopted to facilitate the subsequent excavation, which could not have been so well done if the brickwork had been continued to the bottom, The square timbering, however, will be explained next in order after the present subject.

The brickwork of a nine-foot shaft forms a cylinder of great weight, as it contains 3 tons, 7 cwt. to every yard in depth, and the friction of its circumference against the earth is not sufficient to resist its tendency to slide down, when the ground is removed from under it, to construct the tunnel. It is therefore usual to support the shafts by some means until that portion of the tunnel immediately beneath them, commonly called the shaft length, is completed. The shafts are then permanently connected thereto by a curb of cast-iron or brickwork, as shewn in figs. 1 and 2, plate 5, and thus made secure.

I

The shafts may be secured until they can be permanently united to the tunnel, either by supporting them below, or suspending them from above. At Blechingley the former plan was adopted, and under each shaft was fixed at right angles to the direction of the tunnel, a pair of sills formed of whole balks, 15 inches square, and 34 feet long; and upon them was fixed a square frame of the same kind and scantling of timber, to carry the bottom ring or curb of wood, and the superincumbent shaft. The under side of the sills were placed three feet above the top of the intended brickwork of the arch, that the miners might have plenty of room for their bars, &c., in excavating for the side lengths.

In consequence of the gradient of the railway of this place being lowered, after the shaft-sills were fixed, they necessarily were so much higher above the top of the tunnel: this, however, was attended with but little inconvenience.

The annexed engraving shews the sills and the square frame in detail, with the bottom curb, and shaft resting on them. Fig. 1 is a section of the lower part of the shaft, and fig. 2 a plan shewing the shaft resting on the square frame that is supported by the sills. The shafts being but nine feet in diameter, the sills were scarfed in two places, that they might form three pieces, for convenience in lowering and fixing them. These scarfings, together with the glands that connect the 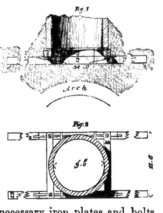 square frame with the sills, and the necessary iron plates and bolts, are shewn in the two figures.

For the insertion of the sills, small headings were driven each way from the shaft, no larger than was sufficient for a man to work in; and when the sills were properly placed, and the square frames attached and screwed down by the glands, and the bolts passing through both the square frame and the sills, the headings were filled with the earth previously excavated, and rammed solid; a good bottom curb was then placed on the frame, and the brickwork made good to underpin the part of the shaft previously constructed, and thus the whole shaft rested on the sills.

In plate 3 the sills and frames are shewn as fixed; at fig. 1 they appear in the direction of the tunnel; and fig. 4, at right angles thereto, *a* the sills, *b* the frame. The sills are also shewn in each of the figures in plate 4, and in figs. 1 and 2, plate 5; the last two shew their appearance and position when the whole was completed.

The stratum of earth through which for the most part the shafts were sunk at Blechingley, was a hard blue bind (or shale) so highly indurated as to be when first exposed like rock, which required to be blasted for the economical working of it. A mass so compact was capable of bearing the weight of the shafts by means of the sills, when the ground beneath was cut away for the shaft length, as shewn at figs. 1 and 2, plate 4. Some assistance however may mostly be obtained by means of props from the bars of the shaft length, and from the projecting ends of the crown bars of the side lengths, as shewn at *a, a, a,* &c. figs. 1 and 2, plate 4. But in all cases where the ground is not a solid or compact mass, when for instance it is loose (or quick) sand as was the case at Saltwood, the use of shaft sills is injurious rather than beneficial, because the ground having no cohesion in itself cannot form a foundation for the sills to rest upon. Under such circumstances they would require to be supported, and thus produce a source of difficulty and danger of no small magnitude: this may be fully understood by reference to plate 4, fig. 2; for if the ground at *b, b,* was loose sand, it would be liable to give way under the sills when the excavation was made for the tunnel.

These considerations led to the omission of the sills for the shafts at Saltwood, and to the suspending them from the surface of the ground by means of hanging rods, which are generally made of bar iron; but there being no suitable material of that kind on the ground, and for the sake of economy, they were constructed of wood, as shewn in plate 2, where both a front and side view of the hangings is given, and their construction rendered plain. A square frame of whole timber under the brickwork was carried by the hangings, and was sufficiently stout to prevent any unequal settlement of the shaft.

The timber of the hangings was larch, of a good quality, being the only available material at hand; and the pieces were scarfed together to obtain the proper length, as shewn in the figures, plate 2.

In most of the shafts the hangings stood the pressure without exhibiting any apparent deficiency of strength; in one or two they appeared weak for the work; and in one case they broke, or rather tore away at the scarfing, but having shewn previous indication of so doing, any casualty was stopped by timely propping.

This apparent weakness and breakage was chiefly attributable to the following cause:—It has been stated that the work was being done through loose sand, the ground having in the first instance been saturated with water to a very great extent; when the water was subsequently drained therefrom, the ground was left in a porous state, and yielded in all directions before any pressure; and where the lower part of the shafts were sunk and square-timbered, a great quantity of what was then a quicksand ran into the shaft, and was removed with the water, leaving large vacuities or caverns in the vicinity, which were unknown to us, and therefore to the great peril of the shaft so situated; for a small amount of lateral or unequal pressure would then throw an unfair strain upon the hangings. This was probably, the cause of the apparent weakness in one or two of the hangings, whilst the others stood sufficiently firm.

The direct cohesion of larch timber, or the weight that a square inch would bear without being torn asunder, as stated in Tredgold's Elementary Principles of Carpentry, page 39, is, according to Rondelet, 10,220 lbs.; and, according to Bevan, 8,900 lbs.; the mean would be 9,560 lbs., which multiplied by the sectional area in inches of the hanging rods, $9'' \times 6'' = 54$ inches, would give 516,240 lbs. as the weight that would tear each rod asunder; but it is further stated, that "the greatest constant load any piece should be allowed to sustain ought not to exceed one fourth of its computed strength; therefore, each such hanging rod should not be loaded with more than 129,060 lbs. or about $57\frac{1}{2}$ tons. The greatest weight of any shaft carried by the two rods was 67 tons, or 33½ tons to each rod, leaving a surplus strength sufficient to have carried fourteen tons more than it was loaded with.

If iron hangings had been used they would have been made of bar iron, of a thickness depending on the weight they would have to carry, which would vary with the different depths of the shafts. In Barlow's "Treatise on the Strength of Materials," page 277, the cohesive strength of good medium iron is assumed upon good data at 56,000 lbs., or 25 tons per square inch; and assuming one-third of that amount as the greatest constant load that should be applied, it would be 8·3 tons per square inch, and as four hanging-rods would be employed, each would have to bear one-fourth of the weight of the shaft, which in the case above quoted would be 16·75 tons Now, an iron rod 1⅝ inch diameter contains 2·07 square inches; which, multiplied by 8·3 tons, gives the weight that such a rod would safely carry, namely, rather more than 17 tons: therefore the diameter of the bar employed to carry such a shaft should not be less than the dimensions above given.

The manner of applying such hanging-rods is shewn in the annexed engraving. Figure 1 is a section of the shaft; and figure 2, a plan at the surface of the ground. Two balks of timber are placed parallel to each other across the shaft, and two other pieces are similarly placed across and at right angles to them, forming a square opening (A fig. 2) to admit the traffic up and down the shaft. The upper end of each rod terminates in a strong well-made screw, and passes through both the stalks and a stout iron plate, and is secured above by a nut; which screws and nuts should 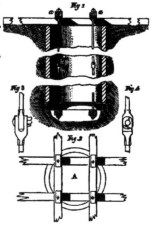 be carefully made, as the security of the whole chiefly depends upon them. Each of the balks should be well bedded on the ground, to give them a good bearing. The bottom of each rod passes also through a balk, two of which carry a strong curb, or square setting, upon which the brickwork of the shaft is constructed.

As it would not be practicable to have the hanging-rods in one piece where there is any great depth of shaft, they must be coupled at different lengths, as may be most convenient. Figures 3 and 4 shew a method of forming these couplings;—the one a face, and the other a side view;—and as the proper lengths of these hanging-rods can be calculated in the first instance, they may be contracted for, and made ready for use immediately that they may be required.

Strong chains would be found a convenient form of hanging; and as they could be of an indefinite length, they would be applicable to any depth of shaft, and lowered as the shaft was extended downwards. Four chains would be required in each shaft, and applied in the same manner as the four iron rods are represented in use in the last figure.

There might be an advantage in the use of chains instead of stout bar iron for the above purpose, as they would be more manageable, and in all probability could be more readily appropriated afterwards to other purposes.

The following particulars may be useful, as connected with this subject :—

TABLE OF THE COHESIVE STRENGTH OF MATERIALS;

Or, the load in pounds that will tear asunder one square inch.

	lbs.
Iron—(good medium)	56,000
Oak—(English)	14,000
Beech	12,000
Ash	17,000
Elm	14,000
Mahogany	12,000
Walnut	8,000
Fir	12,000
Larch	9,560

In the erection of Menai Bridge some trials were made of the strength of ropes used for the hoisting tackle, to get up the main chains. They were as follows:—

	Tons per square inch.
1.—A piece of $5\frac{3}{4}$ inches circumference ... broke with $6\frac{3}{4}$ tons ... ... =	2·56
2.—A piece of $4\frac{1}{4}$ inches circumference, common laid ... broke with $4\frac{1}{10}$ tons ... =	2·54
3.—A piece of $4\frac{1}{4}$ inches circumference, of fine yarn, slack laid ... broke with 6 tons ... =	3·73

Taking the mean of No. 1 and 2 as a standard, it appears that " *good rope will break with a strain of* 2·55 *tons per square inch of section.*" But it ought not to be strained permanently with more than one-third of that,—say three-fourths of a ton. For temporary purposes it might be loaded with half its breaking-strain, or $1\frac{1}{4}$ ton per square inch.

For finding the breaking-strain of ropes, the late Dr. Gregory gave the following rule:—

$$\frac{\text{The girth-square}}{5} = \text{the load in tons that will break the rope.}$$

Which appears to agree well with the experiments at the Menai Bridge:—take for instance the first example, $5\frac{3}{4}$ squared = 33·06, which divided by 5 = 6·61 tons as the breaking strain. The experiment gave a little more, namely, 6·75 tons.

CHAPTER VI.

SHAFT SINKING, CONCLUDED.

EXCAVATING AND SQUARE-TIMBERING THE LOWER PORTION OF THE SHAFTS.

THE underside of the shaft-sills, or of the timber settings carried by the hangings, was three feet from the intended level of the top of the brickwork of the arch; and upon their being made secure, the farther sinking of the shafts through the intended depth of the tunnel was proceeded with. Throughout this space, square settings of timber were placed, at intervals of about six feet, and propped with rough timber from one to the other: the intervals were closely poled, or planked with three-inch deals. The cheapest materials that could be procured for this purpose were six-feet deal-ends. They were placed vertically behind the settings; which kept them tight against the earth behind; with a view to prevent any disturbance of its natural bed,—this being the great object to be aimed at in all the timbering in mining operations: for so long as the earth can be kept undisturbed in situ, the minimum of pressure will be the result; but when once a movement takes place, unequal and uncertain weight is immediately thrown upon the timbers, which too often breaks them, causing a considerable loss of both timber and labour, and frequently attended with danger to the whole of that portion of the excavation, and to the lives of the workmen. It also frequently happens in argillaceous shales, or what the miners call "blue ground," that, upon exposure to the action of the atmosphere, or moisture, the earth will swell, or expand. This was the case at Blechingley; where, occasionally, in the short interval of six feet between the square settings in the shaft, the three-inch planks were bulged or forced out in the middle; which bulging would probably

have gone on until the planks had broken, had this not been prevented by the insertion of an intermediate setting. As this happened
but in a few instances, it would appear that six feet was a proper
distance for the settings from each other; and the ground must be
very bad to require them to be closer: at the same time, it would
seldom be safe or prudent to place them much farther apart; for the
saving would scarcely compensate any risk, as but little more than
the labour in making the settings is lost, for being so soon released
from the shafts the timber can be advantageously employed during
the subsequent works, and the cost of the labour in making them
amounts only to four shillings per setting.

The process of square-timbering, after what has been stated, will
be fully understood by reference to plates 2 and 3. In plate 2, figs.
1 and 2, and in plate 3, fig. 1, the square-timbering is shewn complete. The section, fig. 1, plate 2, is taken across, or at right angles
to the direction of the tunnel; and also shews by the dotted lines
the position that the tunnel would occupy with respect to the shaft.
The opening, or position of the heading is also shewn. Fig. 2 is a
section at right angles to the former, or in the direction of the
tunnel; shewing the arrangement of the timbers in the shaft; and
also a longitudinal section of the heading. Fig. 1, plate 3, is the
same upon a lager scale; and fig. 4, plate 3, shews the upper part of
the square-timbering immediately under the shaft-sills.

The method of framing or putting together the square settings is
shewn by figs. 5 and 6; A is the stretching-timber, which is placed
across the shaft at right angles to the tunnel, as at figs. 1 and 2,
plate 2: B is the side timber placed in the direction of the tunnel,
as in the figs. plate 2; the stretcher, A, has a tenon at each end, to
fix a corresponding mortice in the side timber, B; making when the
four pieces are put together, a clear square opening equal to the
diameter of the shaft above, (in this case, nine feet); but the side
pieces were eighteen inches longer than the stretchers, consequently
their ends projected nine inches beyond the square formed by the
four timbers, and stood out like horns, as shewn at D, figs. 5 and 6.
The use of these horns was to form blocks at the back of the mortices and tenons, to prevent the stretchers from slipping outwards,

K

when the frame was in its place and the earth excavated from behind, during the subsequent excavations for the side-lengths of the
tunnel. In like manner the stretchers were prevented from being
pressed inwards, by chogs, c, figs. 5 and 6, which were spiked to the
side pieces at one end of each stretcher after it was passed into its
place; that part of the side piece being cut away, or sloped, to admit
the tenon to pass into its mortice, which it otherwise would not do ;
as the excavation should not, in the first instance, be made so large
as to admit the side pieces being opened sufficiently wide apart to
allow both tenons to be admitted into their mortices at the same
time. Several ways have been adopted of framing the settings; but
the one above described is probably the best, as affording the greatest
security to the work.

When the ground has been excavated from beneath the shaft-sills
to the proper depth, and the first setting put together, it must be
placed exactly under the shaft, and square with the line of the
tunnel; the earth may then be removed (or rather pared down, if it
will admit of such a process), to allow the insertion of the three-inch
poling-boards, or deal-ends, which should be driven close to each
other, and bedded solid against the earth behind, by packing between
the earth and the boards, if more excavation has been made than
was necessary, or wherever a slip has taken place. The rough
props, E E, &c., plate 2, can then be inserted and wedged tight at
their ends; and, if necessary, spiked to the square timbers, to prevent their moving.

The work to the first square setting will now be completed;
whereupon the excavation downwards may be continued through
another space of six feet, by sinking in the middle of the shaft, and
leaving a projecting bench of earth around, on which the first setting
rests, in the same manner as explained and figured at page 52, when
describing the shaft sinking. When this is done to the proper depth,
the bench is cut away on two sides, for the insertion of the side
pieces of a second setting, which must be placed perpendicularly
under the setting (or settings) already fixed (they being temporarily
propped to keep them from settling during this process). In like
manner the earth is removed for the insertion of the stretching

pieces on the two other sides of the shaft. Some support may be obtained to the upper setting by temporary raking props, and by under-cutting the ground for placing the new setting, and subsequently removing the remainder of the earth to get in the props, E, (plate 2), and the poling boards.

In carrying on operations of this kind, so many new circumstances arise that require different modes of proceeding, even in sinking the same shaft, that it is only possible in a work like this to explain how it may be done, and how it has been done, and to state generally that some judgment is necessary to meet and overcome every difficulty as it arises; and it may be added, watchfulness also, particularly where the ground is not homogeneous, as disasters in tunnel works are seldom rectified at a small cost, and may leave the works in a more or less precarious state.

The foregoing operation, or square timbering, was intended to be carried no lower than the level of the top of the invert of the tunnel, which was also to be the level of the bottom of the heading. The setting marked F, plate 2, occupies this position; but in consequence of meeting with so much water during the shaft sinking, at both the tunnels, the square timbering was carried down one setting, or six feet, below the said level, and thus formed a sump to collect the water, and for the barrels to dip and fill themselves, as they were raised and lowered by the machinery above to draw the water to the surface. The sump is shewn in all the figures representing the square timbering of the shafts.

It may be necessary to explain why in figs. 1 and 2, plate 2, the two settings B' and C, immediately above the heading, are shewn as being so much nearer together than the others. This arose from the use of six-feet deal ends as poling boards, which required that the said timbers should be placed six feet apart from centre to centre. Now, if this six-feet interval had been strictly kept to, the setting C would have been placed directly across the heading D, which it is needless to add, could not be allowed; it was therefore considered better to place the two settings near to each other, and cut one set of polings shorter, rather than, by equally dividing the space above, to have to cut every set of polings to correspond thereto, which would have caused needless waste; or otherwise, the polings must have overlapped each other behind, which would have been troublesome to do, and at the same time not so sound nor workmanlike a job.

In the above manner the work of shaft-sinking was carried on, and satisfactorily completed at Blechingley Tunnel. The quantity of water in the shafts was various; in some it caused delay, in others none worth naming. The jack-rolls, or windlasses, were sufficient for raising both the earth and water, during which time the horse-gins were being made. The price paid for the work, including all labour in drawing earth and water, candles, gunpowder, and all tools, was £4 per yard down. At Saltwood it was commenced at 30s. per yard, and could have been done for that sum, had the sand kept dry, and as easy to get as at first; but when the water appeared in such abundance, of course the work cost much more. It had been the intention to have proceeded at Saltwood in the same manner as at Blechingley, but this was prevented by the great quantity of water, which rendered the ground a complete quicksand. The difficulties met with, and the method of overcoming them, will now be described.

The sinking of the shafts at Saltwood was commenced on the 11th of June, 1842, and was carried on without intermission or difficulty until about the 13th of July; when water began to appear, at a depth varying from sixty to sixty-five feet down. Small barrels were at first used to draw the water, alternately with a skip of earth; but, as the water increased, a second barrel was used at each shaft, and very soon the whole time of the men was taken up in drawing water only; it was therefore evident that the means then employed were inadequate to keep the water under, and enable the work to proceed. It was therefore resolved at once to fix up the horse-gins, and apply much larger water barrels than could be worked by manual labour. The gins were those made for, and used at the works at Blechingley, and now required considerable repairs, which were done as fast as they arrived upon the ground from that place. All the pits, therefore, could not be got to work for some time, but were proceeded with one by one, as the gins could be prepared and fixed.

The horse-gins will be described hereafter; but the large water barrels, with the manner of mounting them for use, is shewn in the annexed engraving. A A are two centres, about which the barrel re- volves when suspended by the large iron bale. The ring B has a double motion, by turning in its socket, and the socket turning in the bale. Strong iron straps pass down the sides, and are crossed underneath the barrel, to strengthen it for carrying its weighty burthen; one of these straps is se- cured to the centre, A, and passes under the barrel to the opposite centre whereby the whole weight is in a measure taken from the bottom of the

barrel, and thrown upon the bale. The centres, A A, are placed below the centre of gravity of the barrel, which therefore will readily tip over, and empty its contents, when raised to the top of the shaft; where a trough is placed to receive and carry off the water. By this method, the barrel need not be landed, but, as soon as it reaches the proper height, the banksman clips the top of the barrel with a hook, and releases the bolt, c, which slides upon the bale, and fits into a socket in the barrel to prevent its revolving until it is required to do so; and as soon as it is thus released, the barrel may be turned over towards the trough, and emptied, without any apparent exertion being required. When emptied, it is placed

upright, and the bolt, c, pressed into its place to keep it erect; and it is again lowered, to fill itself in the sump below. The ring, B, is secured to the end of the gin-rope by means of a shackle shewn in the annexed cut; the rope thus yields and bends as soon as the barrel reaches the bottom, to allow it to roll over, and fill, without any risk of its becoming detached. The same

shackles were used throughout the subsequent works; for when the barrels were not attached to them the large skips for raising the earth—(which will be described, with the gins, &c. in Chap. viii.) —were suspended therefrom, by means of a chain. The rope was $6\frac{1}{4}$ inches in circumference, and cost 50s. per cwt.

The dimensions of the water barrels are given in the engraving; the weight of each was 1 cwt. 2 qrs. 6 lbs.; the ironwork weighed 3 qrs. 20 lbs.; and when full of water the whole weighed 1,310 lbs.: and as they held 100 gallons, the weight of that quantity of water was 1,032 lbs. which gives 10·32 lbs. per gallon, which is about the usually estimated weight of a gallon of water. These determinations have been arrived at, by weighing the parts and the whole, on an excellent weighing machine by Pooley and Son.

When the barrels had been got to work, and the shaft emptied, the sinking was resumed and carried on without intermission. The ground we were excavating, was a dark-coloured sand and clay, nearly black, (but which became lighter as it dried). The quantity of water it held made it of the consistency of soft mud, and as fast as we shovelled it into the skips, the space from whence it was taken was almost instantly filled up again by fresh sand running from the back of the polings around the shaft. In this way we struggled with the work for some time, trying innumerable schemes to counteract the blowing or running of the sand, but to no purpose; for, in several instances, after a fortnight's work we were less advanced than when we began. At length, the following method suggested itself, after the repeated failure of other plans. This was called sumping; and its adoption was attended with success.

The plan was to drive the deal-ends, as if they had been sheet piles, behind the square settings, and remove the earth from the area of the shaft, as they were driven. But, finding it impossible so to do, this operation was preceded by sinking a sump, about six feet square, at the middle of the bottom of the shaft, which was always kept as much lower than the ends of the said piles as was practicable, so that the sump sinking and pile driving were continued together. By this means the water was tended to the sump, and the earth above and around was left in a firmer state; for, it must be mentioned, the water followed us in our descent,—or in other words, the ground was drained to the level of our workings, and left comparatively dry, except in a few cases which were influenced by the upper springs. When this was done, the deal-ends being inserted behind the last square setting, were driven down by beetles (like piles behind a waling timber); at the same time the ground was shovelled from under them to admit of their descent. The earth was kept sufficiently dry for this purpose, by drawing water from the sump as fast as the barrels could be worked, occasionally dis-engaging one of them, to hook on and send up a skip full of earth. As the pile driving proceeded, the sinking of the sump was con-tinued, in order to keep the drainage as much as possible below the lower ends of the piles, and also that the barrels might dip and fill themselves; and when five or six feet were thus gained, another square setting was inserted, as might be found to be necessary; and thus the work was continued to the bottom.

The making of the sump was the difficulty; and this was done as shewn in the annexed engraving, which is a plan or section of the shaft at the level where the sumphing commenced. A, A, are the side pieces, and B, B, the stretching pieces of the square timbers; c, c, c, c, are the upright polings or deal-ends at the back of the square timbers; a, a, &c., are sections of the upright props between the square setting, A, A, B, B, and the one above it; the shaded space represents the bottom of the shaft; and

the space enclosed within the four sides, D, E, F, G, is the sump.

To make the sump, two planks, D, and E, were placed on their edges, parallel to each other, having triangular pieces, or chogs, e, e, e, e, securely spiked to them near their ends; between and at right angles to them two other planks, F and G, were placed, which were kept from being pressed inwards, by the chogs, e, e, &c. When all four planks (or deal-ends) were thus placed, they formed a square frame nine inches deep. The area was then cleared of the sand and water, by two men, the one shovelling the sand into a skip, the

other baling the water into the barrel, at the same time that two
other men with beetles drove down the four planks; for it was
impossible to clear the area within the planks, without at the same
time driving them lower, as the earth ran in so rapidly from behind
as to fill the space immediately. When the planks were driven their
whole depth (nine inches), four other planks were similarly placed
upon them, and all secured together, so as to make a box twice the
depth that a plank is wide, or eighteen inches. The sinking proceeded as before, and when they were down to the level of the
bottom of the shaft, they formed a sump, 18-inches deep. A third
and a fourth set of planks were then placed above them, and lowered
likewise, and so on till a sump several feet deep was attained, for
the barrels to dip and draw off the water, while some progress was
being made with driving the polings or deal-ends behind the settings
as before described. This, together with the sump-sinking was
then continued, as nearly as possibly simultaneously until the required depth of shaft was obtained. Behind every poling-board in
the shaft, and wherever there was any space, a packing of straw was
rammed tight, which had a good effect in preventing the live sand
from running. The extensive use of this kind of packing was of the
most essential service. The state of the work at this time, and
throughout the subsequent driving of the heading, was so injurious
to the health of the workmen, that about one-eighth of the whole
number was under medical treatment for rheumatism, ague, or
dysentery; which, however, proved fatal to but one.

CHAPTER VII.

DRIVING THE HEADINGS; AND EXPERIMENTS UPON HORSE POWER.

WHEN the shafts were completed, the Heading or Adit was commenced. The same dimensions were adopted, and the same settings used, both at Blechingley and Saltwood Tunnels: At the former place the work was done with comparatively but little trouble : the water that appeared passed into the sump at the bottom of the nearest shaft; and, there being no great quantity, the drawing of it to the surface delayed the general work but very little. At Saltwood it was otherwise : there the difficulties continued until the heading was completed through the hill, when the water ran off into the natural water-courses of the country,—previously to which it all had to be drawn to the surface by horse power, as already described. It remains to state a few particulars as to the heading itself.

Figure 1 of the subjoined engraving shews a transverse section, and fig. 2, a longitudinal section of the heading.

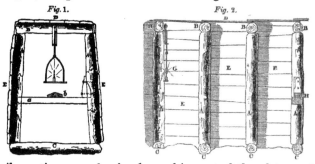

Fig. 1.   *Fig.* 2.

Similar sections may also be observed in most of the plates at the end of the work. The clear dimensions of the heading was 4 feet 8 inches high; 3 feet wide at bottom, and 2 feet 7 inches at the top.

L

Such were the dimensions of the frames or settings, which were made of round larch timber. The caps, B, and the sides, A A, were from five to six inches diameter, and the sills, C, four inches; the sides were tenoned into mortices in the caps and sills; they were placed at intervals of two or three feet apart, according to the character of the ground to be supported. The sides were closed by poling-boards, E, from $\frac{3}{4}$ to one inch thick; and the top, D, with poling not less than one inch in thickness. At Saltwood, the bottom under the sills was also poled quite close, and the whole packed with straw, to prevent the running of the sand, which but for the floor of poling-boards would have blown up from below, and filled the heading. In each of the above figures, the ranging-setting, and candle-holders, described at page 37, are shewn.

The annexed engraving shews the manner of making the necessary excavation, and removing the earth from the heading to the shaft, which was done by means of a skip, upon wheels, which

ran upon a temporary railway. For this purpose, the rails consisted simply of strips of iron, one inch wide and $\frac{1}{8}$ inch thick, screwed down at the edges of long pieces of common fir scantling, 4 inches by 3; which was spiked down to the sills of the square settings, and answered all the purposes required.

The cost of the settings complete, ready for use, delivered on the works, was four shillings and sixpence per setting; and for inch poling-boards, from thirteen shillings and nine pence to sixteen shillings per hundred feet super, also delivered.

The driving and timbering of the heading at Blechingley cost thirty shillings per yard forward, including all labour, candles, gunpowder, and water drawing. At Saltwood, the cost of the labour was the same; the water drawing being an additional expense.

An inspection of the engravings at the end of the work may lead to the remark that the headings were driven at the level of the top of the invert; and a question might arise as to the reason for taking that level, in preference to others. The object in so doing was to keep the drainage of the works at all times free; namely, upon the same level in the finished portions of the work as in the headings at the unfinished parts; and the only interruption thereto arose when the ground was excavated for a new length of brickwork; in which case, the water was carried across this twelve-feet space in a wooden shoot, and what little leaked into the excavation was baled out by the bricklayers' labourers. If the heading had been placed at a lower level, the works would have been always under water to the level of the top of the invert, as soon as any portion of it was completed, and a continued annoyance and expense would have been the result, independently of the greater risk of having unsound work from the effects of the water. Had the heading been driven *altogether below* the invert, and a culvert constructed before the tunnel works had . been begun, a great additional outlay would have been the consequence; and the culvert thus made under ground, in so confined a space, would probably have been badly done, and there would have been an unsound foundation for the invert of the tunnel: moreover, none of the advantages of the heading for the purposes of ventilation, ranging the lines, and levelling, would have been obtained. On the other hand, if the heading had been driven at the level of the *top* of the tunnel, all its advantages in draining the works would have been lost; the ranging by the method of lines along the heading, for setting the ground-moulds, as explained in Chapter iii, would have been less conveniently done, and with greater uncertainty as to its accuracy. If, therefore, there is not a certainty of the ground being quite dry (which is rarely to be expected), such a situation for a heading appears to have but few advantages to recommend it.

HORSE POWER.

During the progress of water-drawing, there appeared an opportunity of obtaining some results as to the power of horses. Having found it desirable to ascertain the amount of labour performed at the various shafts, in order to determine from day to day, not only whether the difficulties were increasing or diminishing, but also correctly to fix the duration of horse-labour at each working,—otherwise there would have been opportunities of deception and misrepresentation,—all the horses being hired (at the price of 7s. per diem). Besides which, it appeared that by keeping a daily register of the work actually performed by the horses in each given time, there would be collected a quantity of facts relative to horse power that might prove useful, in assigning an approximate value to that uncertain co-efficient. The register—which extended from August 25th to October 24th—was kept by my friend and assistant, Mr. P. N. Brockedon, in the manner shewn in the following table, which is a copy of the register of work done at nine shafts, on the 17th of September, 1842; and is sufficiently explanatory of the mode adopted in arriving at the results.

SALTWOOD TUNNEL.—WORK DONE BY THE HORSES IN THE GINS,

For the twenty-four hours ending 6 a. m. September 17th, 1842.

No. of Shaft.	Number raised during 24 hours, of			Average Number per hour, of			Weight in Pounds raised per hour, of			Total Number of Pounds raised the full height.		Height raised, in feet.	Number of Pounds raised One Foot High per Minute.		Time that each Horse work-ed, in Hours.
	Water Barrels.	Large Skips.	Small Skips.	Water Barrels.	Large Skips.	Small Skips.	Water Barrels, each 1310 lbs.	Large Skips, each 1050 lbs.	Small Skips, each 500 lbs.	per Hour.	per Minute.		By two Horses.	By each Horse.	
1	Not in operation.	...	...	...	...	...	...	...	...	...	...	...	...	...	...
2															
3	398	5	100	16·6	0·2	4·2	21,746	210	2,100	24,056	400,9	95	38,086	19,043	4½
4	649	73	...	27·0	3·0	...	35,370	3,150	...	38,520	642,0	110	70,620	35,310	3
5	725	47	...	30·2	2·0	...	39,562	2,100	...	41,662	694,4	108	74,995	37,498	3
6	647	53	...	27·0	2·2	...	35,370	2,310	...	37,680	628,0	107	67,196	33,598	3
7	693	58	...	28·9	2·4	...	37,859	2,520	...	40,379	673,0	103	69,319	34,660	3
8	446	92	...	18·6	3·8	...	24,366	3,990	...	28,356	472,6	101	47,733	23,866	6
9	605	55	...	25·2	2·3	...	33,012	2,415	...	35,427	590,4	100	59,040	29,520	6
10	420	38	...	17·5	1·6	...	22,925	1,680	...	24,605	410,1	100	41,010	20,505	6
11	333	43	...	13·9	1·8	...	18,209	1,890	...	20,099	335,0	95	31,825	15,913	6

Upon the completion of the water-drawing, namely, when the shafts and heading were finished,—the following mean results were obtained as the power of horses working a given number of hours per diem :—

Horses working three hours per diem, mean of 112 results, = 32,943 lbs. raised one foot high in a minute.

Horses working four hours per diem, mean of 4 results, = 37,151 lbs. raised one foot high in a minute.

Horses working four-and-half hours per diem, mean of 12 results, = 27,056 lbs. raised one foot high in a minute.

Horses working six hours per diem, mean of 212 results, = 24,360 lbs. raised one foot high in a minute.

Horses working eight hours per diem, mean of 4 results, = 23,412 lbs. raised one foot high in a minute.

In the determination of the value of horse power from the above results, the three and six-hour experiments alone should be adopted. The other results were more or less objectionable, from a variety of causes over which there could be no control; and are therefore of less practical value.

The following table of estimates of horse power will afford some means of comparison with the above results.

Name.	Pounds raised 1 foot high in a minute.	Hours of Work.	Authority.
Boulton and Watt ...	33,000	8	Robison's Mech. Phil., vol. ii. p. 145.
Tredgold	27,000	8	Tredgold on Railroads, p. 69.
Desagulier	44,000	8	
Ditto	27,500	Not stated.	Dr. Gregory's Mathematics for Practical Men, p. 183.
Sauveur	34,020	8	
Moore, for Society of Arts	21,120	Not stated.	
Smeaton	22,000	Not stated.	

These are higher results than were given by the average of the Saltwood experiments, and more nearly accord with the maximum there obtained : which was as follows.

Horses working 3 hours ... maximum = 47,895 lbs.
Horses working 4 hours 39,694
Horses working 4½ hours 35,300
Horses working 6 hours 36,819
Horses working 8 hours 36,630

But with such high results, or any thing approaching thereto, the horses sunk under the excessive fatigue, and eleven of them died. Nearly one hundred horses were employed,—which were supplied by Mr. Richard Lewis, of Folkestone ;—they were of good quality ; their average height was 15 hands ¼ inch, and their weight about 10½ cwts., and they cost from £20 to £40 each. They had as much corn as they could eat, and were well attended to..

The total quantity of work done by the horses, and its cost, was as under :—

Registered quantity of water drawn 104 feet, the average height, ⎫
　　28,220,800 gallons　　 ...　　　 ...　　　 ...　　　 ... ⎬ = 128,505 tons.
　Ditto,　 ditto,　 earth 3,500 yards ... 1 ton 6 cwt. per yard = 　4,550

　　　　　Total weight drawn to the surface ...　　 ...　　　133,055 tons.

Total cost of horse labour, including a boy to drive each horse ... £1,585 15s. 3d.
　　　 Or 2·85 pence per ton, the average height of 104 feet.

A paper upon the subject of horse power, containing the full particulars of the Saltwood experiments, by the author of this work, was read before the Institution of Civil Engineers, on the evening of March the 14th, 1843.

CHAPTER VIII.

FORM AND DIMENSIONS OF BLECHINGLEY AND SALTWOOD TUNNELS.—
DESCRIPTION OF THE SKIPS, ETC. AND THE HORSE-GINS EMPLOYED
IN THE CONSTRUCTION OF THE WORKS.

THE form and the several dimensions of these Tunnels, as they
were constructed, are shewn at fig. 3, plate 1. The figure is divided
into two parts by a vertical line: the left-hand half shews a transverse section of Blechingley Tunnel; and the right-hand half shews
a similar section of Saltwood Tunnel: the only difference in the two
sections being, that the inverted arch of Saltwood Tunnel had a rise,
or verse sine, of three feet six inches; whereas that of Blechingley
was but three feet. This difference was thought necessary, to ensure
sufficient stability in the former.

The figure of each tunnel, from the springing of the invert on the
one side, over head, to the same point on the opposite side, was
elliptical, and described with arcs of circles whose radii was 21 feet,
15 feet, and 9 feet. The small circles, and the dotted lines, shew
the various centres and the extent of the said arcs. The invert of
Blechingley Tunnel, with a rise of three feet, was struck with
a radius of 22 feet $1\frac{1}{2}$ inch; and that of Saltwood Tunnel, with
a rise of three feet six inches, was struck with a radius of 19 feet
$5\frac{3}{4}$ inches.

The thickness of the brickwork of the invert of the shaft and side
lengths in both tunnels, and for the first leading length at Saltwood
Tunnel, was 2 feet 3 inches; and for the remainder, 18 inches was
considered ample. The side walls and arch at Blechingley Tunnel
were 3 feet thick for the shaft and side lengths: this was, however,
reduced to 1 foot $10\frac{1}{2}$ inches, wherever the ground was sufficiently
sound and undisturbed to admit of it being safely done. At Saltwood
Tunnel, the shaft, side, and first leading lengths were 3 feet, or four

bricks thick; and the remainder of the work 2 feet 3 inches, or three bricks thick. The whole work of both tunnels was set in cement.

The clear height from the upper surface of the rails to the crown of the arch was 21 feet, and the clear width at 5 feet above the said level was 24 feet. The skewbacks and footings were also of brick; and at Blechingley wedge-formed bricks were made for this purpose, as shewn in section at fig. 2, plate 5.

On the right-hand half of the figure are shewn a number of horizontal lines drawn at intervals of 1 foot from the exterior crown of the arch downwards, and the length on each line (in feet and inches) from the centre vertical line to the exterior of the brick work throughout the whole depth of the work is given. Also from the same point (the external crown of the arch) oblique lines are drawn to the extremities of the said horizontal lines, and the length of such oblique lines are given at the outside of the figure on the right hand. By these two sets of lines it will at once be seen that the figure of the tunnel could be laid down independent of the radii given on the left of the figures; their application was, however, to guide the miners in excavating the earth to the correct dimensions: for, as they commenced at the top and worked downwards, they had this guide in taking out sufficient earth to admit the insertion of the tunnel without removing more than was necessary. A plumb-line was hung from the roof in the centre of the tunnel, and upon this line a knot was tied at every 12 inches of its length. A tape, to represent the corresponding line on the working section above spoken of, was stretched horizontally from any one knot to the extent of the excavation, at right angles to the line of tunnel, which determined whether or not sufficient earth had been removed: thus at 8 feet down it required a space of 12 feet 2 inches to be removed on each side of the line, and as a check, another distance to the same point was occasionally measured from the intended crown of the arch which is marked outside of the curve in the figure; at 8 feet down, this oblique distance was 14 feet 6¾ inches.

The dimensions furnished to the miners were those of the outside of the intended brickwork; they had therefore to excavate still farther, to receive the timbers that were required to support the earth.

These extra dimensions were supplied by themselves, at the time; dependant upon the substance of the timbers necessary, together with the requisite allowance for subsidence; as the character of the ground might have pointed out. This extra space for the timber, &c., beyond that required for the brickwork of the tunnel, will be readily understood upon reference to such plates, at the end of the volume, as contain transverse sections of the finished tunnel; of which there are several examples given.

DESCRIPTION OF THE SKIPS, ETC.

The common windlass, or jack-roll, as shewn placed across the top of the shafts, in plate 2, was the only machinery employed for raising the earth and lowering materials, throughout the sinking and heading-driving at Blechingley. The gins (or whims) were not erected until the actual tunnelling operations were commenced; when more powerful means had become necessary. The same windlasses, &c. were subsequently employed in sinking at Saltwood Tunnel; but were there obliged to be laid aside before the sinking was completed, on account of the great flow of water requiring more efficient means to keep it under, as already explained.

As the windlasses were adapted for manual labour, a smaller quantity of earth could only be drawn up at one time than was subsequently raised when horse power was employed. The most convenient sized skip to be used with the windlasses is represented in the annexed engraving. They were made of inch elm, and weighed 84 lbs.; but when filled with wet earth, as that at Saltwood, weighed upon an average 500 lbs., and re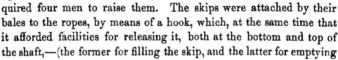quired four men to raise them. The skips were attached by their bales to the ropes, by means of a hook, which, at the same time that it afforded facilities for releasing it, both at the bottom and top of the shaft,—(the former for filling the skip, and the latter for emptying

M

it,)—was contrived so as not to be liberated during its ascent or descent, as such an accident would be likely to prove fatal to the men below. Several schemes were tried for this purpose, and the two found to answer best are shewn in the annexed engraving; wherein both a side and end view of each are given. The left-hand hook is secured from unshipping, by means of an iron pin passed through the return of the hook above a loop dropped over for the purpose: and the right-hand figure shews how the like security was obtained, by a spring acting against a continuation of the  hook, which is thus converted into a ring. This is very similar to the hooks of the hanging-rods represented at page 42. The real dimensions both of the skips and the hooks are given in the above cuts.

At a subsequent period of the work, when the gins were set up, and horse power applied to raise the earth, the larger-sized skips were used. These are represented in the annexed engraving, where the real dimensions are also given. These were not suspended from

the rope by a bale, as were the above described for the smaller skips, but the iron bands terminated in loops, or eyes, one on each of the two opposite sides of the skip, as at A, A. These loops received the hooks of a chain made of $\frac{5}{8}$ inch iron, that was attached to the end of the rope by a shackle, which is represented at page 69, where the method of suspending the water barrels is described. The wheels and axles are attached, for the purpose of running the skips along a temporary railway laid under ground from the several faces of the work to the shaft, and also upon a similar railway, above ground, from the top of the shaft to the tip of the spoil-bank, where the earth raised was deposited. There was no necessity for such an appendage to the smaller skips, because they had to be removed but a few feet, either below or above, and were sufficiently light to be easily shifted by hand: on the contrary, the large skips were too heavy for men to move about, except upon wheels; and the distances they had to be moved continually increased, as the works advanced. They were made of $1\frac{1}{4}$ inch elm; and when empty weighed 140 lbs., and when full of wet earth 1,050 lbs.

DESCRIPTION OF THE HORSE GINS.

The following engravings represent the Horse Gin, (or Whim,) that was employed upon the works. Fig. 1 is a longitudinal view of the gin at work; and fig. 2 shews the same machine as it appears in the other direction, or at right angles thereto. The letters of reference in each apply to the same parts.

Fig. 1.

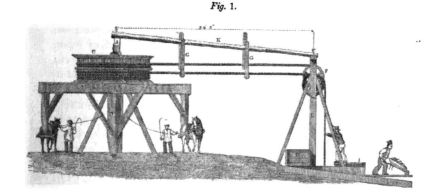

Fig. 2.

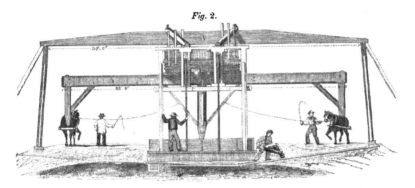

A is the span-beam, with a bearing of 39 feet; 16 inches in
depth in the middle, and 8 inches thick. B is the horse arm, 35 feet
8 inches long; and, as first made for Blechingley works, consisted
of one piece, 12 inches by 7, passed through a mortice in the drum
shaft, D. But finding this mode of construction to be too weak
to stand the strain for a length of time, the horse arms were
subsequently trussed, and consisted of two pieces, each 1 foot 2
inches by 4 inches, notched into and bolted to two opposite sides of
the drum shaft, D; where they measured across, from outside to
outside, 2 feet; and then the truss was gradually narrowed near the
ends, where they measured 1 foot 4 inches, and were blocked apart,
and bolted together through the blocks, between the centre and the
extremities. C is the drum, 9 feet diameter, and 2 feet 6 inches deep,
divided into two parts by a fillet round the middle of the cylindrical
part of the drum, to separate the ascending and descending ropes.
The ropes were prevented from working off the drum, by horns
projecting around the top and bottom, twelve in number, made of
4 by 4-inch stuff, and jutting out about 10 inches; the inner side of
the projection being sloped off. D is the drum shaft, 13 feet 6 inches
long, and 1 foot 2 inches square: to this the horse-arm is securely
bolted at 4 feet from its top, and steadied by stays of 4 by 4-inch
scantling, from its lower part; to these stays the driving boys tied
one end of a small cord, which served as a rein to the horses, as
shewn in the engravings. Above the horse-arm, and resting on it,

the drum was secured by the shaft passing through it, the frame of the drum was made square in the centre for that purpose; the lower frame of the drum was steadied from the shaft at right angles to the horse-arm, by stays, similar to those last described. The shaft rotates on two spindles of 2-inch round iron, one end being squared and turned both ways, at right angles, to secure a firm hold in the shaft. The top spindle worked in a socket formed by straps bolted to the span-beam, and the bottom spindle worked in a bell-metal cup. The top of the shaft had one hoop, 2 inches by $\frac{1}{2}$-inch, driven on after the spindle was fixed; and the bottom of the shaft in like manner had two hoops. The weight of the top spindle was 28 lbs.; the bottom spindle, 25 lbs.; the straps for the top spindle, 18 lbs.; the three hoops, 21 lbs.; and the brass or bell-metal cup, 13 lbs.

The horses were harnessed to shafts attached to a perpendicular piece, 8 inches by 10 inches; its top forming the end of the truss of the horse-arm, from which it was pendant: it was also steadied thereto by means of braces, as shewn in the engravings. The harness shafts were made to revolve on a spindle at the bottom of the perpendicular piece last named, whereby the horses could be turned round, and proceed in the opposite direction, to reverse the action of the machinery; which was necessary, at every ascent and descent of the ropes,—the one ascending as the other descends, and *vice versa*. It would, however, have been more convenient had the machinery of the gin been so contrived that the alternate ascending and descending action of the ropes could have been effected without the necessity of turning the horses round.

The span-beam, A, was supported at each end by a triangular frame—shewn in the last engraving—the base of which was 13 feet long, and was steadied with props. E is the pit frame, placed on each side of the shaft, at a distance of 24 feet 6 inches from the drum shaft. This carries the head-gearing, or frames, in which are mounted the pulley-wheels or sheaves, which were of cast iron, 3 feet in diameter, and weighing 1 cwt. 3 qrs. The spindle upon which the sheave revolved was of wrought iron, 2 inches diameter; and worked in a bell-metal box, or plummer-block, weighing about $3\frac{1}{2}$ lbs. K was a pole, of which there were two, called jackanapes poles, because they carry what are technically called the jackanapes, G, G, whose use was to keep the rope straight in passing from the drum to the head-gearing, and had small friction rollers for it to work upon. These jackanapes were pendulous, and therefore they vibrated or yielded as the ropes moved; which was necessary, because the ropes continually changed their levels as they wound round the drum, or *vice versa.*

H shews the trough into which the water barrels emptied themselves when tilted,—as described at page 69; from which trough the water was passed into proper drains, and was not allowed to soak into the ground. I shews the platform, made to run upon flange wheels, which worked upon rails, whereby it could be drawn over the shaft to cover it when necessary, for greater security in landing the skips full of earth, as they were raised to the surface; the skips were then rolled away, on temporary rails, to the spoil-bank, and emptied of their contents. But, when the wet earth was brought up, at Saltwood Tunnel, during the sinking and water drawing, it was generally emptied into barrows, and wheeled away. This was being done when the sketches for the foregoing engravings were taken, and which accounts for their being shewn therein.

The pit frame, and head-gearing, were made of oak; the frame of the drum of oak, and the covering of elm; the jackanapes poles were of larch; the horse shafts of ash; and the rest of the gin, of Dantzic timber.

For the greater preservation of the ropes, they might be tarred, and payed over with coarse canvas tarred. Some such covering is requisite for economy's sake, as the wear upon the rope is considerable. It may be worth remarking, that the recently invented wire ropes would in all probability be applicable to the purposes now under consideration, not only on account of their apparent greater durability, but to prevent the possibility of wicked persons cutting or otherwise injuring the ropes to cause accidents by their breaking when loaded. Such a circumstance occurred at Balcombe Tunnel, upon the Brighton Railway; where a rope having wilfully been cut, broke at a time when several men in a skip were suspended by it; whereupon they fell to the bottom of the shaft, and one of them was killed. This, unhappily, is not a solitary instance, as the same kind of injury was done to one of the ropes at Blechingley Tunnel, but was fortunately discovered before that it was again put into use. Too great care cannot be exercised where there is so large a body of men congregated together; and some of whom are too apt to indulge vindictive feelings from motives of revenge.

It is not without considerable hesitation, that as Editor to the Second Edition of this very valuable work, I venture to make the few following remarks on the preceding chapter; the more so that many serious and fatal errors have arisen, when the methods laid down have been deviated from, without great care being at the same time bestowed on the work; errors, to remedy which have caused a great loss of time and considerable expense.

As regards the transit instrument and the observatory, it may be observed that where the length of tunnel is not very great, the transit theodolite of 6 inches diameter, as now constructed, may be used instead; nor is an observatory always absolutely required, if a calm day be chosen for the operation of setting out the longitudinal centre-line.

Let the centre-line be carefully ranged out with the instrument by means of stakes, fixing the centres of the shafts, and the stakes required for the spikes mentioned in Chapter III, and also some two or three stakes driven close down into the ground, so that the theodolite may be placed over them; into each of these latter stakes, let a small brass-headed nail be driven, and let these nails be ranged with the utmost care by means of the instrument; these latter centres being thus carefully ranged, their perfect accuracy may be tested by removing the theodolite to these central points, one after another, at the same time examining the correctness of the position of the spikes.

If all things have been properly prepared for doing this, and if all has been kept in perfect readiness for carrying out this operation, it will be evident that no very great time will be required.

Let this be considered as merely preliminary, and on the first calm day let the theodolite be carefully set up over one of the stakes with the nail driven into it, selecting one that will command the best position so as to range backwards and forwards over the whole length of line, and also obtain a view of the two *distant* points that range with the centre-line; this being done, let the *centres* of every stake, those with the spikes on each side of the shafts, as well as those into which the nails have been driven, be all carefully verified.

If this be carefully done, and the centres be found correct and thoroughly in one visual line, as seen through the telescope, there will be no fear but that a perfectly straight line has been obtained; and the lines being strained from spike to spike over the position of the shafts will secure a correct position for the plumb-lines to govern the central line for driving the heading.

As the shafts are sunk and the headings are being set out, it is again necessary to verify the central position of the plumb-line; and this after the earth has been raised round these shafts, so that the banks are too high for the lines hanging from the curbs to be seen with the theodolite; their correct position may at any time be tested by the theodolite in the following manner, and after the observations that have been made by the talented and experienced Author of this work, it will readily be seen that too much care cannot be bestowed in fixing the centres given by these plumb-lines with the utmost caution, inasmuch as it is entirely by these that the central-line of the heading is determined.

Then to test this after the shafts have been sunk to their proper level, let a few light triangles be constructed so that the said plumb-line may be appended from them, and let these be set up approximately over the centre-line; let the lines be stretched from spike to spike, and the plumb-lines from the triangles be made to coincide with them; let the theodolite be set up over one of the brass nails above mentioned, and let it be observed whether the plumb-lines range true with the centre-line set out, when the optical axis of the telescope ranges true with the *distant* objects already mentioned, or with one of them and as many of the little centre discs on the spikes as may be visible from the station when the theodolite is set up.

By these means it will often be found practicable to avoid the necessity of an observatory and transit instrument; in these days of economical railway construction, every item of expenditure has to be carefully guarded against as much as is consistent with safety; where however, the tunnel is of great length, the transit and the observatory should be resorted to for the security of both the contractor and the company, and for the satisfaction of the engineer.

Shafts and Heading.—Frequently the positions of the shafts, instead of being set out on the longitudinal axis of the tunnel, are ranged on a parallel line at a distance of from about 40 to 45 feet. When this is the case, cross-headings are driven from these shafts to the axis of the tunnel in order to drive the main heading. It will at once be perceived that this operation will add considerably to the difficulties of ranging a perfectly straight line under-ground, coinciding with the longitudinal axis of the tunnel, but there are circumstances advantageous to this mode of proceeding; during the construction of the works there is less danger arising from anything falling accidentally down the shafts, the cross-headings form convenient receptacles for the tools or materials, and after the construction of the works, they are so many sanctuaries for the retreat of workmen engaged on repairs or maintenance of permanent way, or for the deposit of tools, if they are at first driven at a sufficient depth.

The bottom of the shafts may also be sunk to a depth of from two to four yards below the level of the lower level of the cross-headings, and a platform laid over this extra depth in continuation of the footway of the heading; it forms a convenient drainage or well for the water, from which it may be pumped up.

When these side-shafts are resorted to, it is evident that the same means are to be resorted to for ranging or setting out as those already mentioned, with the exception that the operations are carried on along the line parallel to the longitudinal axis of the tunnel; the *horizontal* distances between these two lines will require setting out with minute care, and with rods made on purpose; the new line being ranged, distant objects must be set up to determine its position in the most accurate manner possible; a standard measurement of the rods should be established somewhere on the ground, in order that their lengths may be tested when they are afterwards used in the underground workings in order to set out the true horizontal distance between the centre line and the parallel line, from the centre of the shaft at bottom to the longitudinal axis of the tunnel. No small amount of care will also be required in making all these distances truly rectangular with the centre and parallel-lines. The reader will not fail to remark that the Blechingley and the Saltwood Tunnels presented very serious difficulties in their construction, traversing most treacherous strata which yielded vast quantities of water; where in the construction of future tunnels similar conditions obtain, very nearly the same or quite analogous means will have to be observed; greater experience will nevertheless reduce the expense of the works. Fortunately, however, more favourable soils often present themselves to tunnelling operations, and less expensive means may be resorted to for carrying on the works; and as regards the shafts, instead of these being all constructed of brickwork, they may under more favourable conditions be entirely timbered from top to bottom, as it is often done, and constructed square on the plan.

In the angles are placed, *vertically*, longitudinal timbers about 6 feet long, and 9-inch scantling; these vertical pieces, *a*, are maintained by horizontal cross-pieces, *b*, of 9 by 4½ inches, and by two similar pieces, *c*, notched fore and aft to the vertical timbers; the three last pieces are bolted together; behind comes thick planking, *d*, and the whole is tightened up by wedges, *e;* the annexed cut shews this timber-work in plan; as the planking is introduced, care is taken to pack close behind it.

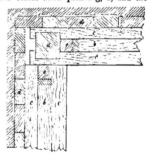

The vertical timbers, A, are scarfed at the end to admit of the next length being secured to them as the operations are carried on, and the scarfed pieces are bolted together. One moment's study will shew that there is great strength in this system, and that it is very easily put together—careful packing behind the planking is an important element in its security.

CHAPTER IX.

CONSTRUCTION OF THE TUNNELS.

THE SIDE LENGTHS.—EXCAVATION AND TIMBERING.

THE excavations for the tunnels at Blechingley and Saltwood were carried on in a similar manner. One description of the general process will therefore suffice; with such occasional particulars of any peculiarity in the circumstances of either, as may have arisen in the course of those works.

The work was commenced by removing some of the polings, or deal ends, from behind the two top settings of the square timbering of the shafts; and driving a narrow heading, about 12 feet long, at the top, and in the middle of the intended tunnel. Where the ground is good, and will stand without much timbering, the top heading (as it is usually called) may have rather large dimensions; but must be limited in this respect where the ground is loose or treacherous. The headings at Blechingley and Saltwood were sufficiently high for a man to stand upright in, and about 3 feet in width. In some of the headings at the former tunnel no poling boards were required in so small an excavation, but at the latter place they were in all cases necessary. No regular system of framing was used, but pieces of poling boards were put up and secured in the best and most convenient manner, wherever the earth shewed symptoms of falling in, but so arranged (where it was possible) as to form part of the subsequent roof of the excavation. The top of this heading was so much above the intended soffit of the arch of the tunnel, as to admit the proposed thickness of the brickwork, and that of the crown bars, packing, and poling boards, together with the allowance of several inches for the settlement of the timber which is certain to take place when more of the excavation is made, and before the brickwork can be inserted to take the weight, and relieve the bars of their burthen. This allowance should never be omitted, for when such settlement takes place, and no room has been previously left for its occurrence, a part or the whole of the crown bars in sinking occupy the position of the intended brickwork; and therefore, in order to insert a tunnel of the required dimensions, the bars and poling boards must be raised to their proper level; which is only to be done piecemeal, by removing the earth over each bar, and then raising them one at a time: this involves considerable labour and care, and no trifling expense.

When the heading is driven, it is widened at the top along one side, to form, as it were, a shelf, upon which a crown bar may be laid lengthways. When this is done, the centre crown bar is placed along the top heading, and supported against the roof, by an upright prop at the remote end, and by resting it on the square timbering of the shaft at the near end; poling boards are then arranged above the two bars to carry the earth. This is shewn in the annexed section of the top heading. A similar excavation or shelf is next made on the other side of the centre crown bar, and a third bar placed thereon, and poling boards inserted above, as in the first instance; a narrow slip of ground is next removed from under the remote ends of the two side crown bars, to the bottom of the heading; and rough props inserted to support them in the same manner that the centre crown bar is supported; their other ends being in like manner supported by the square timbers of the shaft. The earth may next be removed from under the two side bars, which leaves the heading much wider than before.

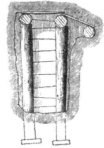

Sometimes, when the top heading is wide enough, two crown bars are inserted and poled above, and the insertion of the side bars (by excavating a shelf to the right and left, as before described,) is then proceeded with, in the manner shewn in the annexed engraving. The bars are kept at the proper distance apart, by inserting five or six struts between every two bars, as shewn in the next engraving, also at s s, &c. in fig. 2 plate 3, and in each plate that contains a transverse section of the timbering of the tunnel. The temporary props, at the remote end of the bars, rest upon flat foot-blocks, to prevent the super-incumbent weight pressing them down. The foot-blocks are either placed at the bottom of the heading, or the ground is dug up to admit of their base standing upon the intended level of the under-side of the top sill. In either case, they are placed far enough outwards to admit thereafter of the sill being placed in front of them.—The dotted portion of the props and foot-blocks, in the above cuts, shews the end of the props so placed below the bottom or floor of the heading: however, it is not always that the ground will allow of this being done in the first instance. The perpendicular face of the work is secured from falling in, by the insertion of poling-boards across it, at the back of the props, as shewn in the last figure.

In the manner above described, bar after bar is inserted to the right and left of the top heading; propped and strutted from the ground and from each other; and the poling-boards inserted both in the roof and against the face of the excavation, the bars being so arranged as to follow nearly the intended figure of the tunnel: or rather, such an arrangement is preserved as will be best suited for the subsequent insertion of the brickwork, as will be hereafter explained.

The annexed engrav-
ing shews a section of
the work in this stage
of progress, which is
technically called "get-
ting in the top."

From what has above
been stated, together
with an examination and
comparison of the en-
gravings, it may be
hoped that the matter has been sufficiently explained to require no further observations.

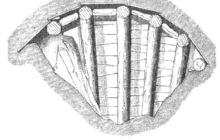

It has been stated above, that the near end of the crown-bars is at first temporarily supported or propped from the square timbers in the shafts; it must however be observed, that, by so doing, a great weight is thrown upon the square timbers in addition to that of the brickwork of the shaft, which is all that it is designed to carry, and in which it is materially assisted by the hanging-rods, or shaft-sills described in the preceding chapters; and, for this reason, the square timbers should be as speedily as possible relieved from the weight of the bars, and whatever pressure of earth they may be sustaining :— this is finally done, when the top sill next the shaft is inserted in its place, by propping every bar therefrom. When the ground is good, there is no danger in temporarily supporting the near ends of the crown-bars from the square timbers; but where it is soft, or yielding, it is unsafe thus to load them; for, under such circumstances, the ground, instead of steadying the square timbers, is liable to give under the pressure; and when once the square timbers get out of the perpendicular they would require no great additional weight to force them in, and the yielding or soft ground which would thus lead to the accident, would follow from behind the shaft, and in all probability bring the shaft down with it.

This was precisely the kind of accident that occurred, in one instance, upon first starting the Saltwood Tunnel. The sand had become extremely porous, and consequently yielding, by the draining of the water therefrom, to a great extent, as described in the preceding chapters. Not only was the earth porous, by the absence of the water, but large spaces (in some instances, complete caverns) had been formed behind and around the square timbers, by the sand having run into the shaft, and been drawn to the surface with the water. Under such circumstances, it was no wonder that half the weight of the first side lengths being thrown upon the square timbers (in consequence of the bars having been propped therefrom), should have caused their downfall, and the destruction of the shaft: the porous nature of the ground, and the existence of the caverns, being unknown at the time. The necessity of relieving the square timbers from the weight of the bars, as speedily as possible, cannot be too strongly impressed upon the reader; and such relief may be obtained to a great extent, if not wholly, by temporarily propping, upon planks or timber laid across the then floor of the excavation, near the shaft, or at the middle of the length, or both, as circumstances or the progress of the work will admit of, during the removal of the earth for, and the insertion of the shaft top sill, D fig. 1, plate 3, which should be got into its place as speedily as possible.

The manner of forming the roof of the excavation by poling-boards over-lapping each other behind the bars, is shewn in the adjoining engraving; where a miner is represented as preparing for the insertion of another bar.

A completely arched roof of timber is constructed, in the manner now de-scribed, and shewn in the upper portion of fig. 2, plate 3, above the sill, A A; but, as at present explained, it is left supported at the face by props resting upon the earth, at the level of the under side of the top sills, each prop standing upon a broad base in the form of a foot-block, to prevent its being pressed into the ground, which would cause the roof to give way: therefore, as soon as the whole of the bars to the intended level of the sills are inserted, or even sooner if possible, the sills themselves should be got into their places. The bars are longer than the first length of brickwork is intended to be; so that, at the face of the excavation, the top sill, A, may be placed in front of the advanced props, and as soon as the sill is placed, another prop may be fixed from the sill to support each bar. These props, therefore, stand in front of the advanced props, which are hid in the section, fig. 2, plate 3, but are shewn in the longitudinal section, fig. 1, plate 3; where A is the top face sill,—F the crown bar,—G the advanced prop, corresponding to those shewn in the cuts at pages 91 and 92, and H the permanent prop. In a similar manner the sill, D, next the shaft, is inserted, and the bars propped therefrom, as I, fig. 1.

The sills (which are called miners' sills, to distinguish them from the centre sills, to be hereafter spoken of), are made of whole balk,

12 or 14 inches square, according to the nature of the ground; or, in other words, the scantling for the timber of the sills must depend upon the weight or pressure they are intended to sustain or resist. Their length was about 36 feet, in two pieces, scarfed together in their middles, and secured with iron plates, bolts, and glands, as shewn at fig. 2, plate 3, where A A is the top sill, and c c the lower sill. The weight of the iron work for the scarfing of each sill was 1 cwt. 2 qrs. 14 lbs.; and the price paid for labour, including sawing, was 4s. 6d. per scarf. Each scarf was five feet in length.

When the sills are well bedded in their places, they should be set level, and the props from them to the bars above, be driven or wedged tight, and secured from the chance of disturbance by driving a peculiarly formed spike, called a *brob*, of wrought iron, (as figured, half the real size, in the margin,) around the ends of the props into the bars and the sills. The brobs are shewn in their places, in figs. 1 and 2, plate 3; and in most of the engravings where their use is required. The tops will now have been completely got in, and the whole weight of the bars, with that of the earth pressing on them, will be supported by the sills, which by their greater length present a large base, and no settlement can take place in any one bar or prop but must equally affect the whole, unless the sills should have been unsoundly bedded in the first instance, or subsequently unequally propped when the earth is removed from below, in the further progress of the work. At this stage of the work the stretchers M M, plate 3, should be inserted, to prevent the sills from being pressed inwards by the earth against the face of the work. The right-hand portion of fig. 1, plate 3, shews the excavation with its timbering ready for the reception of the brick-work, as it appears longitudinally, or in the direction of the tunnel; fig. 2 shews the appearance of the same timbering, as seen at right angles to the former; and by a comparison of the two figures, the corresponding parts in each may readily be distinguished.

When the top sills are in their places, and the roof finished as
above described, the excavation downwards, for the insertion of a
second sill may be proceeded with. So much care is not required
in this part of the work as in getting in the tops. A narrow passage
should first be made along the middle of the length, corresponding
with the top heading before described, and temporary raking props,
κ κ, fig. 1, resting upon foot-blocks, should be placed under the sills,
to carry them till they can be propped from the next lower sill,
after its insertion. The temporary, or, more properly speaking,
advanced props, κ κ, rake outwards, to admit of the sill being
placed and adjusted vertically under, and in all ways parallel to the
one above; whereby when the final props, P P, are inserted between
them, the lower sill will receive its weight in a manner best
calculated to sustain it. For the insertion of each raking prop, a
narrow space is first excavated from under the sill, which leaves
between the places so excavated, a pillar (or pilaster) of earth, that
supports the sill until some of the raking props, κ κ, are secured in
their places. The remaining earth may then gradually be removed,
and other raking props inserted, until sufficient stability is secured to
the sill above.

What has now been stated relates only to the face sill. The
mode of proceeding for the corresponding sill, E, (plate 3, fig. 1,)
adjoining the square timbers of the shaft, is somewhat different,
inasmuch as raking props cannot be in this case applied, excepting
near the extremities, on account of the open space forming the shaft.
The support required for carrying the upper sills and the tops may
be obtained by temporary props raking inwards, placed where they
are not in the way of the insertion of the sills; or if so, they must
then be shifted to another place. And where there is confidence in
the stability of the square timbering of the shaft, a portion of the
weight might be temporarily carried by propping the sill from them,
as shewn at L, fig. 1, plate 3; but this mode of proceeding is best
avoided, unless the circumstances of the case compel the miner to
have recourse thereto, for reasons before explained.

As soon as the upper sills are temporarily secured, and the earth
cleared away from under them to the proper level, the second sills

may be inserted, and the upright or permanent props F' F' be fixed between the bottom and top sills. These props should be set perpendicularly and vertically under the upper ones, which cannot be done unless the sills are truly under each other. When this set of props is inserted and properly secured with brobs, the remainder of the earth, in the length down to the level of the second sill, may be removed; and as the sides are excavated to the required form and dimensions of the tunnel, by hanging a centre line as described at page 80, additional bars, poling-boards and props may be inserted to support the earth. Stretchers also, from sill to sill in the direction of the tunnel, as shewn at M M, &c. plate 3, must be inserted, to prevent the sills from collapsing by the pressure of the earth on either face of the excavation.

In all cases, except where the ground is good, the faces of the excavation, as well as the roof, should be poled behind the advanced or raking props, more or less close, according to the character of the earth; and where it is running sand they should be well packed behind, and at the joints, with straw; which at Saltwood proved a valuable auxiliary.

The section, fig. 2, plate 3, shews the arrangement of all the timbers described between the top sill, A A, and the bottom sill, C C. This arrangement is precisely that followed at Blechingley; but at Saltwood a third sill was used, which was placed immediately over the top of the heading. The second sill was placed midway between the upper and the lower one—see plate 8—which represents a section through the middle of the side lengths at Saltwood, and shews the three sills, and the mode of timbering there adopted; and which in all probability would be suitable for the heaviest ground that occurs in ordinary practice. Such a peculiar situation as that of the Thames Tunnel of course forms an exception; so extraordinary a work required means to be employed that were, in like manner, out of the common way, and could only have been supplied by such a master-mind as his who executed it—Sir M. I. Brunel. When a third sill is used, the mode of excavating between it and the one above is precisely the same as that just described.

The following engraving shews the excavation going forward for

o

the second sill; but it is likewise intended to represent the work in a more advanced state, namely, the construction of the *leading* lengths; and, as such, it will be again referred to in a subsequent

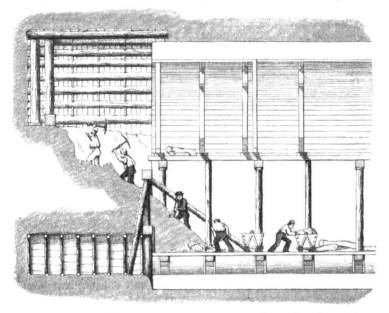

chapter. But, although it does not represent the work quite as it would proceed for a *side* length, yet, by examining and comparing it with what has been stated, some help may be obtained in understanding the present descriptions.

After all the sills required for each length are inserted, the excavating and timbering of the lower portion of the length is very simple. A gullet, or narrow passage, should be excavated down to the level of the skewback of the inverted arch through the length, similar to the narrow passage described at page 96, and which is there represented as corresponding with the top heading previously described. The gullet must then be widened out to the full width of the tunnel; and as the earth is removed from the vicinity of the bottom sills, care must be taken to prop them in the most convenient manner that the state of the works will admit of, whilst the earth is gradually being removed from under them, to admit of upright props being there placed to carry the weight. But temporary propping must first be resorted to, (in this, as in almost every instance of tunnelling operations,) because the permanent (or final) props cannot be got into their places until the ground has been excavated beneath the level of the skewback, in a proper figure for the reception of the inverted arch; which must be the next and last operation of the miner, previous to the bricklayers entering upon the constructive part of the business.

In plate 3, and in most of the engravings at the end of the work, the timbering of the lower portion of the side lengths is shewn: and, after what has now been described, together with an inspection and comparison of the figures in the several engravings, no difficulty can arise in comprehending the whole of it.

Considerable care is necessary in shaping the ground for the inverted arch; for it is as important that it be constructed of as true a figure as the arch overhead is required to be; and which the bricklayers cannot do in a sound and satisfactory manner unless the ground is correctly shaped for its reception.

In the figures at plate 3, and some others, the timbering is shewn as if continued down to the bottom,—at least to the skewback level. This was in all cases necessary at Saltwood, but only in some instances at Blechingley; for there the ground was occasionally so good as to stand, for the short time that it was required to do, without any timber below the level of the bottom sill, as shewn in plate 6.

When the excavation of the side length is complete for the reception of the brickwork, it presents the appearances shewn in section at fig. 2; and the right side of the shaft, at fig. 1, plate 3. Before, however, describing the construction of the side lengths in brickwork, the following particulars, shewing the amount of labour expended in the excavation of the side lengths, both at Blechingley and Saltwood, may be useful and interesting.

NUMBER OF MEN AND HORSES, AND THE NUMBER OF SHIFTS

Employed in the Excavation of the Side Lengths at Blechingley.

Number of Shaft.		Miners.	Labourers.	Horses.	Shifts.
1a	West	111	109	29	26
	East	93	100	22	21
1	West	122	109	28	25
	East	105	100	22	21
2	West	95	106	30	21
	East	66	90	20	15
3	West	122	127	42	29
	East	100	110	27	26
4	West	107	113	33	26
	East	103	96	33	24
5	West	85	80	22	18
	East	81	75	30	18
6	West	96	94	36	24
	East	84	76	27	19
7	West	109	100	30	27
	East	68	59	24	19
8	West	83	83	33	20
	East	65	67	27	15
9	West	94	93	24	20
	East	87	87	28	18
10	West	111	104	31	26
	East	93	88	29	20
11	West	132	119	37	30

Mean of the whole twenty-three lengths :

Number of Miners	...	...	per length	...	96·2
„	Labourers	...	...	... ...	95·0
„	Horses	...	...	... ...	28·9
„	Shifts	...	...	... ...	22·1

Through a misunderstanding on the part of the person who noted the amount of labour expended on the works, the particulars of the first side lengths at Saltwood were not recorded ; but those of six of the second side lengths were noted, and are given in the following table.

NUMBER OF MEN AND HORSES

Employed in excavating six of the second Side Lengths at Saltwood.

Number of Shaft.	Number of Men and Horses employed in driving top Heading.			To top Sill.			To middle Sill.			To bottom Sill.			Invert.			TOTAL.		
	Miners.	Labourers.	Horses.	Miners.	Labourers.	Horses.	Miners.	Labourers.	Horses.	Miners.	Labourers.	Horses.	Miners.	Labourers.	Horses.	Miners.	Labourers.	Horses.
1	3	3	1	14	18	6	21	27	10	16	22	6	9	13	3	63	83	26
2	2	3	1	13	20	6	22	28	8	18	23	6	8	12	3	63	86	24
3	2	4	1	23	34	10	19	25	7	14	17	5	11	13	4	69	93	27
4	2	4	1	21	29	8	18	21	6	14	18	6	11	14	4	66	86	25
5	2	3	1	13	24	8	11	16	4	16	23	6	10	15	4	52	81	23
6	Not worked			...	...	...	...	...	...	...	...	...	...	...	...	...	...	...
7	3	3	1	15	27	8	11	16	4	16	24	6	12	18	6	57	88	25

Mean of the whole six lengths :

Number of Miners	...	...	per length	...	61·6
„ Labourers	...	...	...	...	86·1
„ Horses	...	...	...	...	25·0

As the above statistical observations upon the side lengths at Saltwood were made in so much detail through each stage of the work, it may be useful to draw therefrom the following *average* amount of labour expended upon each portion or subdivision of the work.

AVERAGE AMOUNT OF LABOUR

Expended in each portion of the Excavation and timbering the six Side Lengths at Saltwood Tunnel.

	Getting in the Tops.		Middle Sill.	Bottom Sill.	Invert.	Total.
	Top Heading.	Top Sill.				
Miners...	2·33	16·50	17·00	15·66	10·16	61·65
Labourers	3·33	25·33	22·16	21·16	14·16	86·14
Horses.............	1·00	7·66	6·50	5·83	4·00	24·99

The amount of labour, and hence the cost of getting in the tops,—which is always the most costly part of miners' work, in tunnelling operations,—may be found from the sum of the two first columns of the above table.

By comparing the results of the preceding tables, pages 100 and 101, it appears that less labour was required for the excavation of the side lengths at Saltwood than at Blechingley. This arose from the unexpectedly more favourable character of the ground after it had been so effectually drained by the construction of the headings. The comparison is shewn in the following table; but more extended remarks upon this subject will be made in a subsequent chapter, when there will be an opportunity of making a fairer comparison by means of the results of the leading lengths.

	Blechingley.	Saltwood.	Difference.
Miners.........	96·2	61·6	34·6
Labourers	95·0	86·1	8·9
Horses.........	28·9	25·0	3·9
Shifts	22·1	·	· ·

Upon examining the preceding table for Blechingley, it will be observed that the west side length of each pit (which was the first that was excavated) occupied more time and labour than the east or second excavated length. This may appear remarkable, as it would naturally be expected that the two corresponding lengths in each pit would give a similar result, inasmuch as they were excavated by the same men, and under similar circumstances. It may, however, be accounted for, by supposing that at the first opening of new ground which was understood to be heavy, a greater degree of caution was used by the workmen, and, consequently, their proceedings were slower than when, by completing one length, they had become familiar with the peculiarities of the soil.

When it is stated that the work required 96·2 miners, 95 labourers, &c., it is meant that the labour was equal to 96·2 miners working one day (or shift), or one miner working 96·2 days.

The cost to the contractors for excavating the side lengths at Blechingley would, upon an average, be as follows:—

				£	s.	d.
Miners	96·2 days	...	at 6s. ...	28	17	2
Labourers ...	95·0 ...		at 3s. 6d.	16	12	6
Horses	28·9	...	at 7s. ...	10	2	4
Candles ...	4 dozen ...		at 6s. 6d. ...	1	6	0
Gunpowder ...	1¼ cwt.	...	at 46s. ...	2	17	6
Tools, and sharpening picks, wedges, &c.				1	5	0
Contractors' Superintendence, 22 days, at 7s. ...				7	14	0
Clearing up the work when completed ...		per length		0	5	0
		TOTAL ...	...	£68	19	6

Thus the cost to the contractor averaged £68 : 19 : 6 per length of 12 feet.

In making the engagement with the gangers, or subcontractors, a price per lineal yard, for the side and shaft lengths taken together, was agreed upon, which price was £15, or £60 for each side length; they to find all manual and horse labour, candles, gunpowder, working tools, &c. Now it was well known that at such a price no profit could be derived from the side lengths; as the working expenses would, upon an average, exceed such price, which was proved by the result, as shewn above; but taken together with the shaft lengths, which, at the same time that they were longer than the side lengths, required *much less time and labour to construct*, they yielded a fair amount of profit. When the particulars of the shaft lengths have been given, this subject will be recurred to, for comparison between the actual average cost, and the price paid to the contractors.

When the *leading* lengths were in progress, the miners obtained a bonus, in a charge to the bricklayers of £3 per length for lowering their materials, as bricks, cement, &c., to the underground works; which was done by loading the descending skip, at the time that the earth from the excavation was being raised in the other. This yielded a profit of about £2—the third pound being paid for extra labour in loading the bricks, &c., and the loss of time occasioned to the miners' own work. But during the construction of the *side* and *shaft* lengths no such profit could be obtained, because the excavation was at a total stand, whilst the bricklayers were at work in each of these three lengths; whereas, during the progress of the *leading* work, the bricklayers would be proceeding at one end, from the shaft, whilst the miners would be progressing at the other, and *vice versa*, whereby the earth excavated by the latter could be raised to the surface at the same time, and by the same power, that the materials of the former were lowered.

The observations made of the side lengths at Saltwood, as given at page 91, comprising as they do but one set of side lengths only, are, perhaps, not sufficiently extensive to warrant an investigation of their average cost to the subcontractor, as done above in the case of Blechingley; but if it be found desirable, such cost may be deduced by the aid of the above investigation of the side lengths at Blechingley and of that which will be given in a subsequent chapter upon the leading lengths at Saltwood. The item for the gunpowder must be omitted from such an investigation, as none of that material was used in the last-mentioned work. The quantity of candles consumed, and the charge for superintendence must also be taken in proportion to the comparative quantity of time consumed in executing the work.

CHAPTER X.

CONSTRUCTION OF THE TUNNELS, CONTINUED.

THE SIDE LENGTHS.—BRICKWORK.

THE first thing to be done as soon as the excavation is completed, is to set a ground mould at each end of the length, to guide the bricklayers in constructing the inverted arch to the required form and dimension:—D fig. 2, plate 3, shews a ground mould as set in its place ready for the bricklayers to work. It forms part of what is called a *leading frame*, of which E E are the side walls, or, rather, the moulds to which the side walls are constructed. F and G are stretchers, or cross bars, which connect the parts of the frame together, and keep them at their proper distance apart; and thus altogether these several parts form the leading frame, and, placed upright against the timbered face of the excavation, as shewn at O O, fig. 1, plate 3, guides the bricklayers in carrying on the work.

The inverted arch is built in front of or against the ground mould D, the two ends of which are formed as at L, and constitute the skewback, from which the side walls of the tunnel spring. The points where the curve of each side wall meets that of the invert mould as at *a*, is the part alluded to in Chapter III., page 40, &c. as the "*invert skewback*," when describing the requisite levelling operations. At fig. 2, plate 3, and at plate 8, (which represent the side lengths of the two tunnels under consideration,) it will be observed that the ground moulds are not embedded on the ground, but elevated nine inches above it, and propped in that position, in several places, by bricks laid flatways on each other, and wedged up to the proper level; this was done for all the side lengths at both tunnels, and also for the first leading lengths at Saltwood. By reference to the transverse sections, fig. 3, plate 1, it will be seen

P

that the brickwork of the side and shaft lengths was thicker than in the leading work. This is the usual practice, because these lengths have always more work to do, being liable to be tried with greater strains than the other portions, particularly the side lengths when first constructed, for they then remain a long time before they receive any assistance from the adjoining work.

The ground moulds were made of Dantzic timber, 3 inches thick, and in two parts, scarfed and united in the middle by iron plates and bolts, as shewn at *b*, fig. 2, plate 3. It would have been better to have made these in one piece, but in that form they could not have been got down the shafts into their places; their length being too great to admit of their turning the angle from the shaft into the excavation, through the confined spaces between the timbers; but where shafts of a larger diameter are used such ground moulds may then be got down whole. The stretcher, G, was in one piece, 5 inches by 3 inches, and was made to drop into a socket at each end, formed by iron plates at the skewback, as will be shewn more fully in the engraving on the next page. I I, are two upright pieces, or plumb rules, fitted between the stretcher, G, and the invert mould, D, by mortices and tenons; their use was to give stability to that part of the frame. By their plumb lines the ground mould was set upright, and would, if all its parts had been made very correctly, have determined when the mould was truly level; but it was never depended on for that purpose, as the spirit level, when placed in the length, was both a correct and ready means of so doing. Upon the stretcher, G, and upon the ground mould, the centre line was marked with a saw kirf, to enable it to be placed centrally under the ranging line, when stretched along the heading for this purpose, as explained at pages 38 and 39. In each side length there was of necessity two leading frames used for the starting of the work; after which, one only was required in each length, as the brickwork already constructed answered the purpose of the other. On the right hand side of the shaft, fig. 1, an end view, or section, of the leading frames is shewn, at o o, as set against the back and front of the excavation, ready for the bricklayers.

The side walls of the tunnel were built outside of the moulds,

E E, or between the moulds and the earth; the intervening timber (the bars, &c.) shewn in the engraving fig. 2, were removed as the brickwork was advanced upwards. These wall moulds were also made of Dantzic, 3 inches in thickness, and were fitted on the skewback of the ground mould, by having at their lower end an iron cap, containing mortises that fitted on the corresponding iron tenons, on the plates of the skewback of the ground mould; by this means it could be shipped or unshipped most readily. The subjoined engraving shews this part at large.

Fig. 1 is an eleva-tion, and fig. 2 a plan, of that part of the leading frame which forms the skewback of the inverted arch, or the point where the side walls rise or spring from the invert.

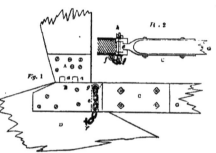

A is the cap before spoken of, on the lower end of the side wall moulds, having mortises at the under side, which drop over two iron tenons, *a a* (shewn by the dotted lines), which rise above a plate screwed down to the invert mould, at its skewback. B, fig. 1, is the skewback plate,—there being a corresponding one at the other side of the invert mould. These plates are so formed as to make the socket into which the end of the stretcher, G, is dropped; its ends being made to correspond thereto, by a cap, c, which is bolted through the wood. At *e*, fig. 2, is shewn, in plan, the end of the stretcher when in the socket above spoken of; *g* is an iron pin passing through the socket, above the end of the stretcher, and retains it in its place. The pin is secured, against working out, by a cotter, *h*, on the other side, and, from being mislaid or lost when not in use, by a chain attached to a staple, at *f*.

The top bar, or upper stretcher, of the leading frame, F, plate 3, served to connect the upper part of the frame, and to keep the top ends of the side wall moulds at the proper distance apart. It was in one piece, 5 inches by 3 inches, and was notched into the side walls, and secured thereto by iron plates, which were bolted to the stretcher, and projected over the moulds on each side. A pin was then passed through the overlapping ends and the moulds, similarly to the pin *g* in the last engraving, and was in like manner secured from being lost, when not in use, by chains attached by staples to the mould.

Above the leading frame, in fig. 2, plate 3, there is a dotted curved line that shews the intended underside of the brickwork, or the position that the tunnel would occupy with respect to the timbering of the excavation; fig. 3 is a cross section of the brick-work of the side length when completed, with the centres and laggins under the arch; and the timbering of the face of the excava-tion, also the crown bars, F F, &c., above the arch, intended to be drawn forward when the leading lengths shall be proceeded with, or left to remain, according to circumstances. Between each of these bars, a pier, *c*, or what is called a packing, is built, as the work pro-ceeds, to relieve the bars from the superincumbent pressure, if it should be deemed advisable to draw them. The left-hand side of the shaft, fig. 1, shews a longitudinal section of the above described length in brickwork.

It has been before stated that in each of the side lengths, two leading frames were required to start the construction of the tunnel. These ground moulds required to be carefully set, and secured in position, so that the centre line marked thereon might coincide with the intended centre of the tunnel, and that they might also be at their proper relative level. For this purpose the level of the skew-back of the invert was adopted in all cases, as before stated. The method of setting these leading frames, or ground moulds, was as follows :—

After being lowered and put together, they were approximately placed against each face of the excavation; the hanging rods were then suspended from above, to obtain the levels, as described in Chapter III., page 40, &c., and the ranging line was stretched along the heading on each side of the shaft, (being passed through two or more holes in the ranging blocks, *b*, described at page 38), representing the intended central line of the tunnel, with which the centre mark of each leading frame should coincide. A spirit level was then set up in a convenient part of the excavation, to determine the levels of the skewbacks by the hanging rods, the bottoms of which were graduated, as at A in the engraving, page 42, similarly to a levelling staff. If this had not been done, a staff must have been held alongside the rod, which would not have been so satisfactory an operation. To determine the levels of the ground moulds, a staff divided like the hanging rods was held thereon, and by directing the level to each, alternately, it could be ascertained if the mould was too high or too low; in either case it was adjusted to the proper level by means of wedges, and, if necessary, more earth was removed from under it. A lighted candle held near the staff, or rod, was sufficient to render the graduations distinctly visible.

In the above manner the ground moulds were set exactly in line and at the proper level; but in addition thereto it was necessary that the plane or face, of the leading frame should be perpendicular, and be at right angles to the centre line, the former was regulated by the plumb lines, I I, fig. 2, plate 3, and the latter was determined as follows :—A nail was driven into some timber at the shaft end of the length, exactly in the centre line, and from thence was measured the distances to the skewback points at each end of the mould. If the two measurements were equal, the mould would be correctly square; but if the distance of the central nail to one end of the mould was greater than the distance to the other end, it was a proof that the mould was not at right angles to the centre line; and, therefore one end was moved inward, or the other end outwards, or both, until the distance from the said central nail, to any assumed point on the mould (as the edge of the skewback) was the same.

This explanation will be better understood by reference to the annexed plan. Let A, B, represent the centre line of the tunnel; C, D, one of the ground moulds; B, a nail (as above described) in the centre line. From this nail the

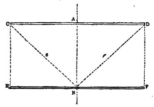

distance B D must be equal to the distance B C, in order that the mould, C D, be at right angles to the line A B,; if it be not so, the mould must be so moved until it answers these conditions, in order that it may be correct.

When one of the moulds is set square, the other may be brought parallel thereto by simply moving it, until its two ends be equidistant from the two ends of the one already fixed. Thus, in the engraving, if E F represent the second ground mould, it must be moved until its centre coincides with the centre line, A B, and the distances of its ends, E and F, are equidistant from the ends C and D of the other mould respectively. When both the ground moulds are set *straight, level, and square*, they may be secured in their places with respect to each other, and prevented from collapsing or expanding, by means of narrow pieces of board nailed from one to the other near to their ends; and also, with respect to their position in the tunnel, by a kind of holdfast, called a dog, (shewn in the annexed cut), driven into the mould and into the adjoining timbers, wherever it may be convenient. This kind of holdfast is very useful in such like operations.

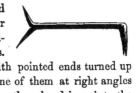

They consist of a piece of round iron, with pointed ends turned up at right angles to their length, and in some of them at right angles to each other: these points, or spurs, can then be driven into the timber in opposite directions. A number of them, of various sizes and degrees of strength, should always be at hand, to be applied as circumstances may require.

During the early proceedings at Blechingley, the first ground moulds were ranged centrally, by suspending two lines down each shaft, from the ranging frame described at page 32, and by moving the mould until the two lines were seen to cut its centre, where a lighted candle was placed. This was but an uncertain and unsatisfactory mode of proceeding, even when all circumstances were favorable; but when, from thick weather, the lines could not be tested above by the transit; or, if tested, could not be depended upon, by reason of high winds forcing them out of the perpendicular;

it became so doubtful, that it would have been imprudent to adjust by them at all,—consequently it occasionally happened that a delay took place in setting the moulds,—and thus a large excavation was left for some time to the strength of the timbers only to carry the earth; which, in unfavourable ground, to say the least of it, was hazardous. This kind of delay led to the fixing of posts securely in the invert, and adjusting to each of them an iron cap, with a central hole through which to pass a line to another central point, fixed at a distance in the heading, whereby the line was passed over each ground mould that would require adjusting; and having thus obtained central as well as level points below, the work could be carried on without delay, or dependance on the contingencies of the weather; and led, at Saltwood, to the contrivance and adoption, in the first instance, of the ranging spikes described and figured at page 35.

It may not be thought unnecessary to recapitulate the points to be attended to in setting the ground moulds.

> 1st—The furthest mould from the shaft should be fixed.
>
> 2nd—It must be upright as determined by the plumb lines attached to it.
>
> 3rd—Its centre must coincide with the centre line of the tunnel, as determined by the ranging lines.
>
> 4th—Each end must be set to the same level; which level is to be derived from bench marks already established below, or from the hanging rods suspended in the shaft for that purpose. In transferring the levels to the skewback, due allowance must be made for the rise or fall of the tunnel (if any) according to the gradient.
>
> 5th—It must be at right angles to the line of tunnel; as determined by measuring from a distant central nail to each end of the mould.
>
> 6th—In the side lengths, the back, or second mould, must be set parallel to the first, by placing it so that its ends may be equidistant from the ends of the mould first set.

When the two ground moulds are set, the miners should trim the ground correctly to receive the brickwork of the inverted arch. This they can do, by following the line of the under side of the moulds.

After the completion of the side lengths, only one leading frame is required to go forward on each side of the shafts, as the work advances; because the brickwork already inserted becomes, as it were, the back mould to be worked from; therefore, as in each shaft four leading frames are required at first, and subsequently only two, the work may be so arranged as to obviate the necessity of making double the number of frames. This may be done by fixing those ground moulds that are next to the timbering of the shafts at a little distance therefrom, by means of wedges, which may be eased and drawn after the brickwork is in, and thus the frame may be set free; for otherwise the brickwork would jam the frame so tightly against the timbers that it would be impossible to set them at liberty until the shaft lengths were excavated, which is never done until both the side lengths are completed.

If, however, it should be considered the more secure method to build tight against the mould and the timbers, to steady them when the ground is heavy, the moulds must remain until released after the completion of the shaft lengths; under which circumstances, additional leading frames must be made, or some of the shafts remain till the side and shaft lengths of others be finished. No moulds are required for the shaft lengths, as the brickwork of the two side lengths must be worked to.

Upon the face of the ground moulds and side walls every course of bricks was distinctly marked (or cut in), to which the bricklayers were required to work. These lines were determined by the size of the bricks to be used; and unless rigidly adhered to, there would have been great irregularity in the courses at the junction of length upon length, which would have been bad in appearance and deficient in strength.

The invert was constructed with concentric half-brick rings, bonded, in each case where the joints became flush, excepting about six feet on each side from the springing at the angle of the skewback, which was constructed in English bond. At Blechingley the bricks for the skewback were made of a suitable shape, but at Saltwood the common bricks were cut for that purpose. In some instances stone has been employed, whereby the skewback is made in one piece for lengths of about three feet; but this method was considered too expensive for the tunnels under consideration.

When the invert and skewback were completed, the side wall moulds were set up, by simply placing them in the situation shewn in fig. 2, and at o o, fig. 1, plate 3; and fixing the cross bar or stretcher F, at the top; thus completing the leading frame, which was then set upright by the plumb lines, K K, and also central, by suspending from e a plumb bob, d, which should hang vertical over the centre line, as marked on the lower stretcher, G; it was then secured in its position by dogs driven into it and some of the props and sills, and by pieces of board wherever they could be advantageously attached.

The brickwork of the side walls to the springing of the arch was then constructed in English bond, with neatly-drawn joints. As the brickwork advanced, the bars that had supported the earth were removed, together with as much of the poling boards as could be got out. At Blechingley the work was built solid against the earth wherever there was more space than was necessary for the insertion of the intended thickness of the brickwork; but at Saltwood, such vacuities were rammed solid as the work advanced. Whichever of these plans may be adopted, it is of great importance that it should be carefully executed; it should therefore, in all cases, be well attended to.

When the walls were up springing high, the centres were set for turning the arch. The form, and all the particulars relative to them will be reserved for a separate chapter, and it is therefore only necessary in this place to refer to fig. 3, plate 3, for their appearance when set, and of the brickwork of the side lengths completed. The left-hand portion of fig. 1 is a longitudinal section of the same work.

The brickwork of the arch consisted of a series of concentric half-brick rings; and where the joints became flush and straight, a heading, or bonding course, was inserted throughout the length of the arch; and care was taken that the bricklayers should preserve the true form of the arch in every ring of which it was composed.

Q

The manner of removing the bars which supported
the earth, as the upper part of the side walls and the
arch was advanced, is shewn in the adjoining cut;
where the brickwork, A, is represented as brought up
as far as the bar, B; and, before it can be advanced
any further, the bar must be removed.   The brickwork
is built close up to the poling board, C, which retains the earth, and
also overlaps the front of the poling, D; it is therefore clear that the
brickwork would hold both the polings in their places, and the bar
which before held them might be removed without danger; for,
even when the ground is bad, no movement could well take place in
it before the brickwork would be further advanced, and render the
poling board, D, still more secure.   In no case should the bars be
removed until the brickwork is ready to supply its place, without
delay.   In this way each of the side bars were removed, one by one,
and laid aside to be used in the further progress of the work.   A
considerable number of poling boards were necessarily built in,
where the ground was not good, and their removal could not be
effected without risk.

The brickwork of the arch was brought up equally on each side,
towards the crown, until it assumed the shape shewn in fig. 3,
plate 7; where A A is the brickwork of the arch.   At this stage of
the work, the laggins, c c, which are rabbeted on the top of their
inner edge, are placed on the centres.   In these rabbets, cross
laggins, d, about 18 inches wide, are placed, one at a time, beginning
at one end of the length.   A bricklayer then (standing with his
head and shoulders between the two sides of the brickwork, A A)
keys in the arch over the first short cross or keying-in laggin; which
done, he places a second cross laggin in the rabbets of the long
laggins, c c, and in like manner keys in that portion of the arch; he
then places a third cross laggin, and keys in, as before, retreating
backwards along the narrow space between A A, as his work advances,
until the whole of the length is completely keyed in.   These keying-
in laggins are also represented in the
annexed cut,—which is a section through
the laggins,—where A is the top of a

'centre rib; *b b*, ordinary laggins, which were battens, 12 feet long,
.6 inches wide, and 3 inches thick; *c c*, are the rabbeted laggins
extending the whole of the 12-feet length; and *d*, one of the
keying-in laggins fitting in the rabbets.

Fig. 4, plate 7, is a longitudinal section, shewing the same work,
and also the manner in which the end of a length is left, with
respect to the timber props and bars. F is the projecting end of a
bar, to be drawn forward for the next length; G is the back prop,
and H the permanent prop, as described at page 94. The whole of
the brickwork was set in Roman cement.

Upon an inspection of fig. 3, plate 3, seven crown bars appear
to have been built in, and counterforts or packings of brickwork, *c c*,
constructed between each two of them. In all cases these bars were
closed in above the arch, and remained there until the shaft lengths
were completed, because they could not have been previously drawn;
and no good purpose could be answered by so doing; on the contrary, the disturbing of them might have endangered the rest of the
work. When, however, the first leading length was excavated, these
bars were, in some instances, drawn forward, to be used for the
crown of such length. The manner of doing this will be explained
in a subsequent chapter, when describing the leading work. But in
every case where symptoms of much pressure were present, it was
considered prudent to leave the bars of the side lengths wholly
undisturbed; as it was better to lose them than run any risk by
their removal.

The bricklayers should be closely watched during the whole
time they are at work, to see that they do it in a sound and satisfactory manner; as upon their labours the future stability of the
tunnel depends. In no part should more attention be paid than
while they are constructing the invert, for as much depends upon its
strength as upon the arch above; and too frequently this is considered
by the workmen as of minor importance; and as it is more out of
sight when done than other portions of the work, it has a corresponding chance of being slurred over. In many cases, the low price
that the men are paid for their task work leads to their hurrying it
over to make up their wages, and in other cases, where they have

been well paid, unless they have been looked after, the chance of making greater gains has been their inducement to slight their work. The system of sub-letting the work, and then again sub-letting it in detail, wherever it is practised, invariably has an injurious effect upon the soundness of the construction.

The price paid for the brickwork at Blechingley was £4 10s. per rod ; which included the setting and moving forward the centres,— all tools, candles, and the lowering of the materials from the surface to the underground works, procuring water where required, &c.

TABLE SHEWING THE PROGRESS OF THE BRICKWORK

In the construction of the Side Lengths at Blechingley Tunnel.

Number of Shaft.		Ground Mould set.	Side Walls Springing high.	Arch Keyed in.
1a	West	March 25	March 27	March 31
	East	April 14	April 17	April 22
1	West	March 13	March 16	March 19
	East	April 2	April 5	April 8
2	West	April 9	April 4	April 7
	East	April 19	April 21	April 24
3	West	March 1	March 4	March 7
	East	March 22	March 24	March 26
4	West	March 12	March 14	March 17
	East	March 31	April 3	April 5
5	West	April 29	May 1	May 4
	East	May 14	May 15	May 18
6	West	April 14	April 16	April 18
	East	April 29	May 1	May 4
7	West	May 6	May 8	May 11
	East	May 22	May 24	May 26
8	West	May 12	May 14	May 17
	East	May 25	May 27	May 29
9	West	April 14	April 16	April 18
	East	April 30	May 1	May 4
10	West	April 7	April 9	April 12
	East	April 24	April 26	April 28
11	West	March 22	March 24	March 27

From the preceding table the following mean results are obtained:

Shaft.	Time occupied in the construction of the Invert and Side Walls.	Time occupied in setting the Centres and turning the Arch.	Total Time occupied in constructing a Length.
	Days.	Days.	Days.
1a	2·5	4·5	7·0
1	3·0	3·0	6·0
2	2·0	3·0	5·0
3	2·5	2·5	5·0
4	2·5	2·5	5·0
5	1·5	3·0	4·5
6	2·0	2·5	4·5
7	2·0	2·5	4·5
8	2·0	2·5	4·5
9	1·5	2·5	4·0
10	2·0	2·5	4.5
11	2·0	3·0	5·0

Mean result of the whole table:

	Days.
Time occupied in the construction of the Invert and Side Walls ...	2·13
Time occupied in setting the Centres, and turning the Arch ...	2·83
Total time occupied in constructing a Side Length ...	4·96

The number of men and horses employed in the above work was not registered.

For reasons stated at page 100, the particulars of the brickwork of the first side lengths at Saltwood were not recorded; but the following table gives all the requisite information respecting the second side lengths.

TABLE SHEWING THE TIME AND FORCE
Employed in the construction of ten Side Lengths at Saltwood Tunnel.

West Side Lengths. Number of the Shaft.	Ground Moulds set.	Number of Men and Horses on the Invert.			Springing high.	Number of Men and Horses on the Side Walls.			Keyed in.	Number of Men and Horses on the Arch.			Total Number of Men and Horses completing the Length.		
		Bricklayers.	Labourers.	Horses.		Bricklayers.	Labourers.	Horses.		Bricklayers.	Labourers.	Horses.	Bricklayers.	Labourers.	Horses.
1	Nov.23	10	25	4½	Nov.26	10	25	4½	Dec. 1	18	36	9	38	86	18
2	23	10	25	4½	26	12½	26½	6¼	1	18	40	9	40½	91½	19¾
3	25	8	22	4	29	10	20	4½	2	18	42	9	36	84	17½
4	26	8	22	4	29	10	22	4½	2	18	42	9	36	86	17½
5	23	10	24	4½	25	10	20	4½	Nov.30	14	32	7	34	76	16
6	Not worked	...	...	...	...	...	...	...	...	...	...	...	...	...	...
7	23	11	23	4¾	26	11	23	4¾	30	18	42	9	40	88	18½
8	24	14	28	6	28	9	21	4¼	Dec. 2	14	32	7	37	81	17½
9	21	8	20	4	23	8	20	4	Nov.26	14	32	7	30	72	15
10	Not worked	...	...	...	...	...	...	...	...	...	...	...	...	...	...
11	24	8	20	4	28	8	20	4	Dec. 3	14	33	7	30	73	15
12	25	12	26	6	29	10	22	4½	3	18	42	9	40	90	19½

From the preceding table the following mean results are obtained:

	Days.
Time occupied in the construction of the Invert and Side Walls ...	3·2
Time occupied in setting the Centres and turning the Arch ...	4·1
Total time occupied in constructing a Side Length	7.3

MEAN OF THE FORCE EMPLOYED.

	In constructing			
	Invert.	Side Walls.	Arch.	Total in Length.
Bricklayers ...	9·90	9·85	16·40	36·15
Labourers	23·50	21·95	37·30	82·75
Horses	4·63	4·60	6·20	17·43

By comparing the results of the tables at pages 116 and 118, it will be seen that more time was taken by the bricklayers in the construction of the side lengths at Saltwood than was employed, for the similar work, at Blechingley : the former occupying 7·3 days, and the latter but 4·96 days.

CHAPTER XI.

CONSTRUCTION OF THE TUNNELS, CONTINUED.
THE SHAFT LENGTHS.—EXCAVATION AND BRICKWORK.

EXCAVATION.

UPON the completion of the second side length, the excavation of the earth from under the shaft, for the insertion of the shaft length of brickwork, should be proceeded with, as rapidly as circumstances and sufficient care for the safety of the shaft, will admit of. This, like the other portions of the excavation, is commenced at the top, and continued downwards. As the whole space beneath the shaft, to the full width of the tunnel, must be cleared away, it is evident that the square timbers, which had hitherto assisted in carrying the shaft, must be removed also: their removal leaves the shaft to be supported, or suspended, wholly by the shaft sills, as at Blechingley: or by the hanging rods, as at Saltwood; or by whatever other means may have been resorted to for that purpose. It is therefore incumbent on the miner to provide, as speedily as possible, other means of assisting to resist the downward tendency of the shaft, until it can be securely connected with the crown of the tunnel, by means of a curb of brick or cast iron. The only support that can be given thereto, (without encumbering the space where the men are at work,) is by propping from the projecting ends of the crown bars of the side lengths, and from the upper bars of the shaft length, as they are inserted; and in this way ample strength may be obtained for the purpose.

The mode of timbering a shaft length is very simple; and consists in placing the bars with their ends resting on the back of the arch of the two side lengths already completed, and poling behind them, to secure the earth from moving. Plate 4 represents a shaft length, as completed ready for the bricklayers;—fig. 1 is a longitudinal section, and fig. 2 a tranverse section through the middle of the length; and, consequently, each section is taken through the centre of the shaft. Fig. 1 shews the two finished side lengths, A A', and the wide space between them (or the shaft length) supported by the timbers, as above described. B B B, &c. are the bars, whose ends are passed behind the brickwork of the lengths A and A', whereby they are secured in their places. s s s, &c. are the short stretchers between the bars, which keep them steady, and connect them for mutual support. The poling boards are also shown in place. c c, are the crown bars of the side length, the ends of which project beyond their own brickwork, and thus supply a base, from which to prop the sills, or curb, that carries the shaft. At Blechingley, the under sides of the shaft sills were thus supported ; and at Saltwood, in the absence of such sills, the suspended square frame, or setting, upon which the wooden curb temporarily rested, was supported. a a, shows two of the props thus carried by the ends of the crown bars (c) of the side length. In addition to this, more help may be obtained, by propping from the upper bars of the shaft length, as shown at a' a', fig. 1, which are props supported by the upper bar, B'. In this way the shafts were supported, both at Blechingley and Saltwood, until the brickwork could be completed to take the weight of the shaft.

The tranverse section, fig. 2, plate 4, being taken through the middle of the length, cuts through the bars, B B, &c., and shows an end view of the brickwork of the side length, A A, which is left in toothings to be united with that of the shaft length. The same letters of reference in the two sections refer to corresponding parts of the work.

During the excavation for the shaft length, the centres that are under the side lengths, A and A', together with the timbers connected therewith, should remain undisturbed; but the miners' sills (D and E, fig. 1, plate 3,) which were against the shaft, must be removed, with the earth that they abutted against; and in doing this, the stretchers, M M, (plate 3) between the miners' sills, must also be removed. But before the removal of either the sills or the stretchers, it is necessary to secure the timbers of the face of the excavation from the possibility of their being forced inwards by the pressure of the earth behind (especially if it should, as at Blechingley, expand). This security is to be obtained by fixing two raking props, D D, plate 4, against the lower miners' sill; the upper end of the prop being cut in the form of a bird's mouth, to receive the angle of the sill; and strongly hooped, to prevent its splitting; the lower end must be firmly bedded, and wedged into a hole made for that purpose in the brickwork of the inverted arch.

Figs. 1 and 2, plate 4, shew the raking props in place, of which there were always two against the lower sill at each face; and, at Blechingley, there was occasionally a necessity for two others to be set against the upper sill, and which are shewn in plates 6 and 7. At Saltwood it was generally found sufficient to use one pair of rakers against the middle sill only; but in all cases, after the completion of the side and shaft lengths, and in advancing the leading work, four rakers were invariably used, at both the tunnels, for the two upper sills, at every face of the excavation.

Nothing more need be added, as to the mode of doing this part of the work; it remains therefore to shew the time and force employed as given in Chapter IX., page 100, for the side lengths.

NUMBER OF MEN AND HORSES, AND THE NUMBER OF SHIFTS,
Employed in the Excavation of the Shaft Lengths at Blechingley.

Number of Shaft.	Miners.	Labourers.	Horses.	Shifts.
1a	56	58	15	14
1	52	41	14	12
2	42	45	14	11
3	55	57	15	12
4	48	46	16	11
5	38	42	16	11
6	37	39	14	10
7	38	43	13	10
8	40	40	15	11
9	43	40	15	10
10	55	53	18	12
11	Not	noted.	...	...

R

Mean of the whole eleven Lengths :

Number of Miners	...	...	...	...	45·8
,, Labourers	...	...	...	...	45·8
,, Horses	...	...	...	...	15·0
,, Shifts ...	...	...	...	...	11·3

From the preceding results, the average cost to the ganger, or subcontractor, may be deduced, in the same manner as done for the side lengths at page 103 ; thus—

					£	s.	d.
Miners ...	45·8 days	...	at 6s.	...	13	14	9
Labourers ...	45·8	...	at 3s. 6d.	...	8	0	4
Horses ...	15·0	...	at 7s.	...	5	5	0
Candles ...	2 dozen ...	...	at 6s. 6d.	...	0	13	0
Gunpowder...	1 cwt.	...	at 46s.	...	2	6	0
Tools, and sharpening picks, wedges, &c.	...	...	...	...	1	5	0
Contractors' superintendence, 11·3 days	...	at 7s.	...		3	19	1
Clearing up the work when completed	...	per length	...		0	5	0
TOTAL	...	...			£35	8	2

At page 103 it has been shewn, that the excavation of the side lengths alone, at £15 per lineal yard, would have been a losing concern ; but it was there stated, that taken together with the shaft lengths a fair profit was obtained ; this may now be shewn as follows :—

COST OF EXCAVATION :

Two Side Lengths, each £68 19s. 6d.	=	£137	19	0
Shaft Length, as above ...	...	=	35	8	2
Total Cost to the Ganger	=	£173	7	2

The side lengths were each 12 feet in length, and the shaft lengths averaged 14 feet, making 38 feet for the three lengths ; which, at £15 per yard, or £5 per foot, gave £190 to be received by the ganger for the work, which cost upon an average, £173 7s. 2d. ; leaving a clear profit of £16 12s. 10d., or at the rate of about 8½ per cent.

The following table will shew the amount of labour, &c., con-sumed in the excavation of six shaft lengths at Saltwood :—

NUMBER OF MEN AND HORSES
Employed in the Excavation of six of the Shaft Lengths at Saltwood.

Number of Shaft.	Miners.	Labourers.	Horses.
1	49	57	20
2	51	58	20
3	51	60	21
4	52	59	20
5	48	56	19
6	Not worked.		...
7	52	59	18

Mean of the whole six Lengths:

Number of Miners	...	...	...	50·5
„ Labourers	...	...	...	58·2
„ Horses	...	...	...	19·7

By comparing the above table with that at page 121, it will be seen, that more labour was expended upon the shaft lengths at Saltwood than was required for the similar work at Blechingley; and that such a result is the reverse of that obtained by comparing the labour of the side lengths, as at page 102; and also is contrary to the result of a comparison of the leading work, as will appear in the next chapter. The greater amount of labour required for the shaft lengths at Saltwood must have arisen from the circumstance, that greater caution, and expenditure of time, was there required, in removing the timbers about the shaft; in timbering the length; and in making all solid, where the running of the sand, during the shaft sinking, had left large vacuities.

BRICKWORK.

After what has been stated in Chapter X., but little remains to be said of the brickwork of the shaft lengths. The sumps under the shafts were carefully filled solid with dry earth, or with concrete, according to circumstances, and upon this the inverted arch was constructed.

When the side walls were built, the arch was turned upon four centre ribs, *a a a a*, fig. 1, plate 5, in which the mode of executing the brickwork of the shaft length is shewn. The three ribs under each of the side lengths were left in their places, undisturbed, until the shaft length, and also the first leading length, were completed. At Blechingley there were ten centre ribs used in every shaft; five in each direction; so that when the four ribs were lowered for turning the shaft length, they were not again raised to the surface, (except for the purpose of repairs,) until the work was completed. These four ribs, together with the three under each side length, made the required ten; of which five were advanced, as the work proceeded, in one direction, and the other five in the opposite direction.

It has sometimes been the practice, for the sake of economy, to use but six centre ribs in a shaft, instead of ten; and for the purpose of turning the shaft length, to remove the two back ribs, *b b*, fig. 1, plate 5, from each of the side lengths, and place them under the shaft length, in the position *a a*, &c.; then to re-adjust the two ribs, *c c*, nearest the edge of the shaft, to take the ends of the laggins for the shaft length to be turned upon. By this practice the side lengths are left without any assistance, at a time when they are least of all capable of bearing any great strain; and while thus standing alone, and unsupported, they have to carry, not only the weight of earth that is fairly their due, but also the weight or pressure of the earth upon the bars of the shaft length; and in addition thereto, have to sustain the weight of the shaft itself.

The safest practice is in the use of the greater number of ribs; and those were arranged as shewn at fig. 1, plate 5. The three ribs under each side length remaining undisturbed, were a great assistance to the brickwork in resisting any great pressure. And further, the two ribs, $b\,b$, on each side, were continued in their places until the first leading length each way was also completed; for as three ribs only were required for each leading length, two of the four ribs, $a\,a$, were moved forward each way from under the shaft to the advanced work, and the back rib, c, was all that was taken from (or disturbed in,) the side length to make the three ribs required for turning the arch of the first leading length. And furthermore, the ribs, $b\,b$, were still continued in their places, undisturbed, until the side walls of a second leading length were constructed, and ready for the centres; which walls acted as buttresses against the new work when the two ribs, $b\,b$, were removed; and were well calculated to resist any tendency to derangement in the work.

In the foregoing manner the side lengths were well sustained during the whole time they were exposed to more than their fair share of pressure; and the same mode of advancing the work, with five centre ribs each way from the shafts, was carried on throughout the tunnel at Blechingley, without the least accident, or undue settlement, in the arch or side walls.

To determine upon the adoption of the use of either ten or six centre ribs, depends upon whether or not it be considered, that the extra expense attending the former method, more than counter-balances the risk incurred by using six ribs only; the risk not being confined to the construction of the shaft length, but lasting until the work is finished. And it is not improbable that the cost of setting to rights one broken length, would exceed the whole extra cost of the centres, laggins, &c., for any one shaft; independent of the doubt (to say the least of it) that attends all work done where there exists any chance of failure.

When the arch is turned the shaft is permanently connected therewith by a curb, either of brickwork, or cast iron made and put together in segments. The former was used in the works now under consideration, and is shewn at figs. 1 and 2, plate 5. For these curbs the bricks were purposely made of the required shape; the angular bricks, where the shaft joins the soffit of the arch, were made large and rounded off, as shewn in the engraving; these bull-

nosed bricks, (as they were called by the workmen,) not only gave
the work a better appearance when finished, but the arris being thus
taken off, it did not injure the gin rope, in its subsequent ascending
and descending; neither was the curb so liable to sustain injury by
the striking of the skips. This subject will be enlarged upon in a
future chapter, and a description, with an engraving, of the iron
curbs, will be given.

The annexed table shews the amount of labour required in constructing the curb at each shaft of Blechingley Tunnel.

NUMBER OF MEN AND HORSES

Employed in constructing the Brick Curbs to the Shafts at Blechingley Tunnel.

Number of Shaft.	Bricklayers.	Labourers.	Horses.
1a	67·7	59·0	20·0
1	72·7	93·5	18·3
2	59·5	72·5	17·0
3	60·0	74·3	22·5
4	64·7	78·3	21·3
5	66·7	65·0	18·0
6	73·7	93·0	21·3
7	77·7	82·0	22·0
8	68·7	101·0	16·0
9	64·7	65·7	17·3
10	60·0	69·3	17·3

Mean of the whole eleven Shafts:

Bricklayers	...	...	...	66·9
Labourers	...	...	...	...	77·6
Horses	...	...	...	...	19·2

The five bars nearest to the crown of the shaft length, were built
in, and left,—as shewn in the transverse section, fig. 2, plate 5. In
this plate the manner of finishing the shaft, where it joins the arch
of the tunnel, and of underpinning the shaft sills, at Blechingley,
are fully shewn.

At E, fig. 2, is shewn the manner in which the skewback of the
invert of the tunnel was constructed, as regards the form and
arrangement of the bricks.

CHAPTER XII.

CONSTRUCTION OF THE TUNNEL, CONCLUDED.
THE LEADING AND THE JUNCTION LENGTHS.—EXCAVATION AND BRICKWORK.

THE LEADING LENGTHS.

WHEN the shaft length is completed,—the curb inserted,—and the brickwork made good to the shaft,—the centre ribs and the laggins may be removed from beneath it; but those under the side lengths should not be disturbed until after the first length forward is completed, for prudential reasons, explained when treating of setting the centres for the shaft lengths, at page 124, and which will again be alluded to when the centres, and method of using them, is described in a subsequent chapter.

The side lengths at Blechingley and Saltwood were 12 feet long, and so situated as to leave between them, upon an average, 14 feet for the shaft length; the three lengths making together 38 feet of tunnel under every shaft, from which to carry on the work in both directions. When this portion of the work is done, the difficulties of the tunnel may be said to be over; as the subsequent proceedings are comparatively straightforward and safe: at all events, there can be but few natural difficulties that cannot be foreseen, and consequently their effects provided for, or guarded against; unless by injudicious proceedings, or absolute carelessness, difficulties and dangers arise which otherwise would not have existed.

Previously to commencing the leading lengths, it is requisite to construct a platform over the invert of the lengths already completed, as shown at P P, plate 9: and which platform must be continued each way, as the work advances. It is made of planks, laid on sleepers, or transverse timbers, placed across the invert, so as to leave a free channel for the water to pass along the invert, to be drained off through the heading; or, in cases where the water is not abundant, it may hence be conducted to a proper receptacle or sump, convenient for the workmen to use it in mixing their cement or mortar; for where there is no water in the tunnel, the conveyance of that material to the shafts for the bricklayers' use, forms a considerable item of expenditure. This was partly the case at Blechingley; the water, which was in abundance at first, diminished in quantity as the work advanced, and towards the last, (except at the west end) the land springs appeared to have been drained nearly dry.

The more immediate use of the platform is to make a uniform plane, on which to lay down a temporary double line of rails, to run the wheeled skips (described at page 82) upon, from the face of the excavation to the shafts, and the empty skips back again; one line being used as a going, and the other as a returning road. The rails, at Blechingley, consisted of pieces of quartering, with hoop iron fastened along the inner edge of the upper surface, for the wheels to run upon, and were before named at page 74: the roads so formed were frequently out of order, and in all probability it would have been better, in the first instance, to have provided some light wrought-iron railway bars for the purpose, which although much more expensive in their first cost would have saved a deal of annoyance afterwards. The platform is shewn in the engraving page 98, where one man is represented as pushing a loaded skip towards the shaft, and another filling an empty skip with the debris cast down by the men above, who are excavating for the insertion of the second sill of a leading length.

The process of driving a leading length is nearly the same as that described for a side length; with this difference, that the bars in that case have to be propped and supported, at both their ends, whereas, in the leading work, they only require such assistance at their remote end, or against the face of the excavation; the near or back end of each bar, being left to rest behind, or upon the brickwork of the arch already turned.

The work is commenced by getting in the top, in the manner described as for the side length. A top heading is driven in the middle line of the tunnel, for the insertion of the crown bars, and is then widened out to the right and left, and the bars inserted one by one down to the level of the top sill. It will be remembered, that the crown bars for the leading lengths were described as left above the brickwork of the side length, to be drawn out (of their cells) to form the roof of the leading work, length after length; by which means, the same bars travel along the roof to the next junction, unless by accident any of them get broken, or stick fast in some part of their journey; whereupon they are generally built in, and left. It is, however, not the safest of practice to draw the crown bars from the side lengths at all, but to build them in, and leave them, unless the ground is very good, when their removal would be attended with perfect safety. They were mostly left in at Blechingley; as their value was of trifling importance, compared with any risk to the security of the work that carried the shafts, through disturbing the earth thereabouts; for, although the space from whence each bar is drawn is, professedly, rammed solid with earth, by a man standing at the end where it is drawn, using a long-handled punner,—yet, however well and carefully this may be done, it would, in most cases, be better that the bars were built in, than that the surrounding earth should be in any degree disturbed: and, too often, if the men are not watched, they will omit the ramming altogether, as their neglect cannot be detected afterwards.

The next engraving represents the process of drawing the crown bars, whether from over the side or any subsequent leading length. The top heading is shewn as having been already driven, and one bar, A, drawn forwards, and its advanced end resting upon the shelf of earth, B, preparatory to its being propped; the ground is also shewn as ready for another bar, C, which the men are drawing from over the brickwork of the last turned length, D; the leading centre rib, E, is shewn in section under the brickwork, also an end view, F, of the top sill, and the upper end of its raking prop, G.

The drawing of the bars can mostly be accomplished with crow bars, used as levers, as shewn in the next engraving, which brings

s

them forward by little and little, till the larger portion of them is advanced, and then they come out easily enough; but if, during their confinement above the brickwork, any particular settlement has taken place, the bars will frequently be jammed in extremely tight; the only way then to release them is by the use of one or more screw-jacks placed horizontally against the arch, and lashing chains passed over these and also round the projecting ends of the bars, when upon working the screws, the bars are released. If however, the resistance is too great to be overcome in this manner, the bars are left and built in; for where a settlement has been so great as to cause such an effect, it would probably be unwise to draw the bars at all, were it even possible to do so, lest a movement be given to the earth that would be liable to produce results far more costly than the value of a few bars.

When the bars are drawn, great care should be taken, as before stated, that the space from whence they are removed is packed and rammed solid with earth; for the danger of leaving an empty space above the arch is too obvious to need any remarks. It is also of importance that attention be paid to the amount of sinking that takes place in the top of each length whilst standing in timber, in order that the leading ends of the bars for the succeeding lengths may be raised sufficiently high above their required level to allow for their sinking, before the arch is turned.

It will be easily understood that were so large an excavation is entirely supported by timbers, in comparatively small pieces, resting upon and pressing against each other, without being one piece of frame-work well braced together, that there must be some general or particular settlement in its parts; and this, as might be expected, takes place very much in the roof. If, therefore, the bars are raised sufficiently high above their required level, and such settlement takes place, they will merely descend to their proper places; and if the amount of sinking that is thus provided for does not occur, it is but of little consequence, as the space can be packed solid, and a more correct allowance for it be made in future. If, however, sufficient allowance has not been made, the bars will sink into the space that must be occupied by the brickwork of the arch; in which case, when the arch is brought up to that place the bars must be raised, by excavating above them by little and little, which is attended with inconvenience, and additional expense. The arch forms the gauge to direct the miners in placing the bars. This subject has been named before, in connexion with the side lengths,— page 90,—but its importance in every stage of the work will be sufficient excuse for referring to it again.

The engraving at page 98 is a section of the tunnel, with a leading length in progress. The last two completed lengths are left resting on five centre ribs, supported by their props. The whole of the timbering of the top of the new length is represented as complete down to the first sill; and the excavation is proceeding for the insertion of a second sill, which is shewn as standing in its last place against the toothing end of the finished length, supported by its props; from whence it will be shifted forward, as soon as the excavation is ready for its reception.

The whole of the operation of timbering and bricking the leading work is shewn in plates 6 and 7. Fig. 1, plate 6, is a section through the timbering at the line, A B, fig. 2, and shews the face of the excavation, with its sills, bars, props, &c. &c., with a leading frame, o, set ready for the bricklayers: fig. 2 is a longitudinal section of the same thing. It will be observed that the timbering is not shewn to extend lower than the bottom sill, except in the central part, R, around the heading of the tunnel. This is the manner in which a large portion of Blechingley Tunnel was done; the ground being sufficiently good for the lower parts to stand without assistance, during the short time it was exposed, until the brickwork was inserted; but this in no case applied to the upper parts, which had a much longer time to stand before it could be permanently closed in. It was always the rule, never to let a length wait for the bricklayers, but to have them and their materials ready to proceed immediately that the excavators had done, and the ground mould set: thus the lower part of the length being the last exposed, and the first closed in, stood but a short time without assistance.

The leaving the lower part of the excavation without being timbered was not general throughout the tunnel, and in no case around the heading, where the ground was always loose, having been previously disturbed when the heading was driven. The method and extent of timbering about the heading is shewn at R, fig. 1, plate 6. At Saltwood Tunnel, as well as in some parts of that at Blechingley, the timbering was continued to the bottom, as it must be in all loose ground—and is shewn in plate 8, and several other engravings.

The timbering of the side and leading lengths are nearly similar to each other, with this chief difference; that in the former several stretchers were fixed from sill to sill, horizontally, along the length, as described at page 95, to keep the back and forward sill from collapsing, or the face of the excavation being pressed inwards; but in the leading lengths, no back sills are required, as the completed brickwork makes all secure over head, at that end of the excavation; in order therefore to prevent the pressure of the earth forcing the sills and timber of the face inwards, and bringing destruction upon the whole length, two raking props are applied to each sill, notched at the upper end to fit its angle, and wedged at the lower end in a hole purposely left, or made, in the invert; and thus the thrust of the face is resisted. An inspection of each figure representing the leading work will make the explanation perfectly clear. D D and

D′ D′, in each of the figures in plate 6, are the raking props. These raking props have also been previously alluded to at page 133.

A longitudinal section, shewing the brickwork of a length completed, is shewn at figure 1, plate 7; and figure 2 is a transverse section taken across the said length, or on the line A B, fig 1.

After all that has been before stated respecting the manner of excavating and constructing the side lengths, it would be but a recapitulation to go through the details of that of the leading work; the *modus operandi*, in both cases, being so similar that an inspection of the plates 6 and 7, after reading what has before been stated, must show to every intelligent reader the whole business.

Before closing this part of the subject it should be observed, that where the forward ends of the bars, in the leading lengths, abut against the face of the excavation, they should be well chogged, or rather, tightly wedged, against the earth, allowing the bar no interval or room to play in the direction of its length. This is especially necessary where the ground is loose, as sand; or yielding, as soft clay; for as the tendency of all leading work is to settle, or press forwards, in the direction that the work is being driven, the earth in front of the said bars is liable to yield when the pressure is great, and if it does yield the whole length goes forward that much, and is liable to drag the last turned length with it; and thereby cause a fracture in the brickwork where it joined the preceding length.

This kind of accident occurred at Saltwood Tunnel, where the negligent workmen had even left a space between the ends of the bars, and the face of the work, which caused the lengths so circumstanced to go forward, and drag with them the last length of brickwork, breaking it away from the preceding work, and brought so much weight thereon, as not only to break the brickwork, but also to break the bars themselves; none of which accidents occurred after that care was taken that the ends of the timbers were closely abutted (by wedges if necessary) to the face of the work. It must here be again impressed upon the practical man, the necessity of always keeping the work tight against the earth, to prevent the possibility of its moving; and it should be an invariable rule, never to leave a vacuity behind the work.

The quantity of timber required for any tunnel work will depend upon the character of the ground. In the works at Blechingley, which have been described, two miners' sills only were used; the ground at that place was *neither good*, nor, taken as a whole, *very bad;* if it had been a little heavier, or in other words, have pressed more heavily on the work, three sills would have been used in each face instead of two. The extensive use of timber is to be avoided as much as possible where it can be safely omitted, because it increases the cost of the works, not only by the price of the timber, but for additional labour in inserting and removing it. In no place should be practised a penny-wise economy in the use of materials, as that frequently results in pounds of subsequent expenditure. Experience and judgment will decide as to the extent of timbering necessary.

When three sills are employed, as at Saltwood Tunnel, the mode of operation need not be altered from that already set forth; and an inspection of plate 8, which represents the timbering adopted at Saltwood Tunnel, with what has already been stated, will give all the information necessary for the working with three sills; A, B, C, are the three miners' sills; *a a*, &c. &c. show the position of the stretchers, this being a side length; but in other respects, it is the same as was adopted for the leading work, except the omission of the raking props.

Where the circumstances are such that the ground will stand with but little or no timbering, as is mostly the case with rock and chalk, the operation of tunnelling is of the simplest character. The only thing necessary to guard against is the first displacement of the strata; which can generally be prevented with very slight timbering, judiciously placed; if this is not watched, and done in time, a slip of the rock will frequently bring in so much as to leave a great cavern, which must be filled solid behind the work to make it secure from future danger. The annexed engraving shows the manner of timbering in constructing tunnels of this description; and is similar to that adopted by Mr. Wright, in the construction of the Abbot's Cliff Tunnel, which was made through the lower chalk between Folkestone and Dover. The sides are first excavated, leaving a pillar in the middle, which serves as a base to prop the roof from, and also to support the centres for turning the arch when the side walls are up; the pillar may then be cut away. But where an invert is to be inserted this mode of proceeding is inconvenient, and requires great care in propping the arch, during the construction of the invert and the side walls  for the underpinning of the said arch; because the centre pillar that was left to carry the props and the centres must be removed, before the invert can be commenced; and that must be completed before the side walls can be constructed. This method has been practised to a limited extent, namely, excavating and constructing the tunnel from the top down to the springing; or, in other words, constructing the arch first, then excavating the lower parts, and, by constructing the invert and side walls, underpin the arch. Such a method of proceeding is better suited for rock and chalk tunnels, when no invert is required, than for heavy ground where an invert is indispensable; in which latter case considerable risk attends the operation. It was however accomplished very successfully by Mr. Daniel Frazer, in a large portion of the Martello Tunnel, which he constructed near Folkestone, for the contractors, Messrs. Grissel and Peto, through the junction of the chalk, upper green sand, and the gault; which latter stratum appeared in the lower part of a portion of the tunnel.

Having now described the *modus operandi* of proceeding with the leading lengths, it will be necessary to give the particulars of the amount of labour consumed, and the cost of the work.

NUMBER OF MEN AND HORSES, AND THE NUMBER OF SHIFTS

Employed in the Excavation of the Leading Lengths at Blechingley Tunnel.

Number of Shaft.	Number of Lengths constructed from each Shaft.	Average Number in each Shaft of			
		Miners.	Labourers.	Horses.	Shifts.
1a	21	52·1	71·7	15·1	14·3
1	23	42·8	53·9	12·5	11·8
2	24	47·3	65·8	14·7	9·0
3	24	53·1	67·8	16·8	9·5
4	22	51·7	78·4	18·4	10·1
5	16	51·3	79·1	17·4	9·6
6	19	52·4	72·4	18·2	11·0
7	19	48·5	63·5	14·8	8·8
8	20	39·5	54·3	12·4	6·9
9	20	51·6	70·4	16·5	9·4
10	21	64·5	74·8	18·5	10·5
11	11	88·8	112·7	30·8	25·4

Mean of the whole two hundred and forty lengths :

Number of Miners	...	..	per length	...	52·2
„ Labourers	...	...	...	...	70·1
„ Horses	...	...	...	...	16·5
„ Shifts ...	..	...	...	...	10·8

The number of workmen employed in a length will vary in different parts of the work as it advances. In driving the top heading, one or two miners and one labourer only can be employed; then, in getting in the tops and before they have commenced drawing the earth, the numbers will be about three or four miners and three labourers; and when the earth is being drawn to the surface, the greatest force can be put on, which amounts to about five miners, three labourers attending upon them, one hooker-on, and four banksmen.

Upon inspecting the preceding table it will be seen that No. 11 shaft employed the greatest force, and took the most time; this arose from the ground being so very heavy where it was shallow, near the entrance of the tunnel, as described in the general particulars of the tunnel, at pages 7 and 8.

It may also be observed, that the amount of labour for the leading work is much less than that for the side lengths, as given at page 101, although it may, at first sight, appear that as the lengths were nearly of similar dimensions such a difference ought not to exist; but it must be remembered, that at the first starting of all works a deal of time is lost in preparation, &c.; and, on this occasion, all the timber had to be lowered down, and prepared for the side lengths, which was not the case afterwards.

Previously to letting the labour, by contract, to the respective gangers, (each of whom was to have no more than one shaft), the following estimate was made of the probable average expense of executing the work of the leading lengths.

ESTIMATE OF THE COST OF EXCAVATING

The Leading Lengths at Blechingley Tunnel.

		£	s.	d.
Driving top heading, and taking out timber, 4 yards ... at 15s.		3	0	0
Four miners getting in tops ... 4 shifts (extra) ... 6s.		4	16	0
Remainder of length, or drawing the earth; say 8 shifts for the following force :—				
4 miners ... ... 6s. ...	1 4 0			
1 mongrel ... 4s. 6d. ...	0 4 6			
3 fillers ... ... 3s. 6d. ...	0 10 6			
4 banksmen ... 3s. 6d. ...	0 14 0			
1 hooker-on ... ... 3s. . .	0 3 0			
2 horses and drivers 7s. ...	0 14 0			
Cost of labour, per shift ...	3 10 0			
which multiplied by 8 shifts = ...		28	0	0
Powder and candles		3	0	0
Tools, &c.		1	5	0
Superintendence ... 8 shifts ... ... 7s. ...		2	16	0
Total for each Length		£42	17	0

Which, divided by 4, gives £10 14s. 3d. per lineal yard.

The price contracted to be paid to the gangers, for the excavation, varied from £10 10s. upwards, per lineal yard, according to the supposed character of the ground, at the different shafts. The average was about £11. Some portion of the work yielded a deal of water, and was extremely heavy; which brought up the average, as above stated.

T

By means of the table at page 136, the average cost to the contractor may be ascertained, and compared with the contract price and the estimate given above.

AVERAGE COST OF EXCAVATING

The Leading Lengths at Blechingley.

			£	s.	d.
Miners	52·2 days ...	at 6s. ...	15	13	2
Labourers ...	70·1 ...	at 3s. 6d.	12	5	4
Horses and drivers	16·5 ...	at 7s. ...	5	15	6
Candles ...	3 dozen ...	at 6s. 6d.	0	19	6
Gunpowder ...	1 cwt. ...	... ...	2	6	0
Tools, and sharpening picks, wedges, &c.			1	5	0
Contractors' superintendence, 10·8 days, at 7s.			3	15	7
Clearing up the work when completed ...	per length		0	5	0
		TOTAL	£42	5	1

Which, divided by 4, gives £10 11s. 3d. per lineal yard.

Thus the cost to the contractor averaged £42 5s. 1d. per length, and the amount he received averaged £44 :—leaving a profit of £1 14s. 11d. To this may be added £3, which the bricklayers paid to the miner, per length, for lowering bricks, cement, and sand, down the shaft, at the same time that they were raising the earth to the surface ; this, however, would require one or two additional labourers, to load the bricks, &c. into the skips, and would take one pound away from the three, leaving £2 to be added to the above £1 14s. 11d.; making together £3 14s. 11d. as the profit, per length : being at the rate of about 8½ per cent.

The above is about the average result of the works at Blechingley; —in some cases the work did not cost so much money to execute it, and in some it cost more. There were several cases in which the contract price would not cover the outlay, and the gangers at such shafts gave up the work. Upon the whole, the above statement appears to be a fair representation of the cost of executing the work at Blechingley.

The following statement comprises a similar investigation of the works at Saltwood.

AVERAGE NUMBER of MEN and HORSES, and the NUMBER of SHIFTS
Employed in the Excavation of the Leading Lengths at Saltwood Tunnel.

Number of Shaft.	Number of Lengths constructed from each Shaft.	Miners.	Labourers.	Horses.	Number of Shifts.		
					Before drawing earth.	While drawing earth.	Total.
1	17	45·3	50·3	12·8	3.3	7·2	10·5
2	18	42·1	50·1	15·3	2·5	7·2	9·7
3	20	39·2	54·3	15·3	2·7	7·1	9·8
4	18	36·5	54·7	14·4	3·2	6·6	9·8
5	22	33·3	50·6	13·2	2·7	6·1	8·8
6	Not worked.	...	...	...	...	...	...
7	20	34·3	53·6	14·0	2·5	6·5	9·0
8	19	34·6	50·6	13·4	2·7	6·2	8·9
9	20	34·6	53·3	13·8	2·9	6·2	9·1
10	17	34·3	55·2	14·3	2·8	6·3	9·1
11	15	35·3	57·1	14·8	3·1	6·6	9·7
12	6	34·7	56·3	14·5	3·1	6·4	9·5

Mean of the whole one hundred and ninety-two Lengths:

Number of Miners ...	...	...	...	per length	36·8
„ Labourers ...	...	...	...	...	53·0
„ Horses	...	...	...	...	14·1
„ Shifts	...	...	...	...	9·4

The preceding table shews the average amount of labour and
time expended in the construction of the leading lengths at Saltwood
Tunnel. The cost thereof would be approximately as follows:

AVERAGE COST OF EXCAVATING
The Leading Lengths at Saltwood.

				£	s.	d.
Miners	36·8 days ...	at 5s. ...		9	4	0
Labourers ...	53	at 3s. 3d.		8	12	3
Horses and drivers	14·1 ...	at 7s. ...		4	18	8
Candles ...	2¼ dozen	at 6s. 6d.		0	16	3
Tools, and sharpening picks, wedges, &c.				1	0	0
Contractors' superintendence, 9 days, at 7s.				3	3	0
Clearing up the work when completed ...	per length			0	5	0
	TOTAL			£27	19	2

Which, divided by 4, gives £6 19s. 9d. per lineal yard.

The price of the miners and labourers' wages are put above at 5s. and 3s. 3d. per diem; whereas, for the same class of men, in the investigation of the cost of Blechingley they were reckoned at 6s. and 3s. 6d. The fact was as stated, in each case; for wages were higher at the time Blechingley Tunnel was in hand than they were two years later, when Saltwood Tunnel was constructed; because that so many similar works were proceeding at the time the former tunnel was made, which led to a demand for men who were accustomed to the work; whereas, two years later no such demand for workmen existed.

At Saltwood, the contractors supplied the horse power themselves, and let the manual labour, only, to the sub-contractors.

A small portion of the Saltwood Tunnel was driven from the open cutting at the west end, as the excavation was completed at a sufficiently early time for the purpose. Seven lengths of tunnel were

constructed from the intended face, or entrance to the tunnel, when a junction was effected with the workings that had proceeded in the opposite direction from No. 1 shaft.

The mode of proceeding with the work from an open cutting is the same as for the ordinary leading lengths previously described; but considerable care is required in timbering the face of the excavation, before the driving of the tunnel is commenced, to prevent its falling in, and causing inconvenience and expense.

The method of timbering the face of the excavation at the intended western entrance to Saltwood Tunnel, is shewn in the engraving on the page opposite.

When the works were in this stage of progress, the temporary railway that had previously been used in making the open cutting, was advanced into the end of the tunnel, as the work progressed. The earth that was excavated by the miners, was filled into wagons, and removed to a distant spoil-bank. The timbering, and the construction, were in all other respects the same as for the sub-excavation.

The following table shews the average amount of labour and time expended in excavating and timbering each 12-feet length, from the open end, as above described.

Number of	Lengths	...	...	...	...	...	7
,,	Miners	..	...	...	...		34·3
,,	Labourers		...	...	...	...	40
,,	Horses	...	...	...	...		11
,,	Shifts	...	...	...	...	...	9·7

By placing the mean results of the tables in pages 136 and 139 in juxta-position, it will be seen how much more labour was required in the excavation of the leading work at Blechingley, than was necessary at Saltwood. This could not arise from any difference in the skill of the workmen, as a large portion of the same men were employed in both cases; and all the gangers, or sub-contractors at Blechingley were the most experienced men that could be found: the difference arose from the varied character of the ground in the two cases. At Blechingley it was a strong blue clay, highly indurated into a hard shale or bind requiring the aid of gunpowder to get it and when exposed to moisture, or the air, it swelled and afterwards slaked; there was also some water to contend with. At Saltwood after that the preliminary works were completed, which had drained off the water most effectually, the ground was a dry sand, except at the level of the invert, where but little trouble was experienced from the water, as the heading afforded so excellent a means of letting it off. Admitting, therefore, that equal skill was employed in both cases, the following table will shew the amount of labour required for excavating the leading work, in the two kinds of earth, and may serve as a useful guide in future operations.

	Blue Shale.	Dry Sand.	Difference in favour of Sand.	Approximate Ratio of the Sand to the Shale.	
Miners............	52·2	36·8	15·4	0·7	
Labourers	70·1	53·0	17·1	0·7	mean,
Horses............	16·5	14·1	2·4	0·8	0·8
Shifts	10·8	9·4	1·4	0·9	

Or the amount of labour and time required to excavate for tunnelling through dry sand, may be said to be approximately eight-tenths of that required to do the same work in blue clay or bind.

BRICKLAYERS' WORK FOR THE LEADING LENGTHS.

Upon the subject of the Bricklayers' Work but little need be stated in addition to the particulars given in Chapter X., where the construction of the side lengths is described; except to give the following table of the work done at Blechingley. The reason that there is so little variation in the amount of labour and time required for the brickwork arises from the fact that the circumstances are nearly always alike; this kind of work not being subject to such vicissitudes as that of the miners.

AVERAGE TIME TAKEN BY THE BRICKLAYERS

To turn twelve feet Leading Lengths at Blechingley Tunnel.

Number of Shaft.	Number of Lengths.	From setting the Ground Mould to the springing of the Arch.	From the springing of the Arch to Keying-in.	TOTAL.	Force employed.	
					Bricklayers.	Labourers.
		Days.	Days.	Days.		
1a	17	2·3	2·6	4·9	4	7
1	20	2·2	2·9	5·1	4	7
2	23	1·9	2·4	4·3	4	7
3	23	1·8	2·2	4·0	4	7
4	21	1·9	2·4	4·3	4	7
5	15	1·7	2·5	4·2	4	7
6	18	1·8	2·2	4·0	4	7
7	17	1·7	2·5	4·2	4	7
8	19	1·6	2·1	3·7	4	7
9	20	2·0	2·5	4·5	4	7
10	20	1·7	2·1	3·8	4	7
11	12	2·0	3·0	5·0	4	7

Mean result of the preceding table:

	Days.
Time occupied in the construction of the Invert and Side Walls ...	1·88
Time occupied in setting the Centres, and turning the Arch ...	2·42
Total time occupied in constructing a leading Length	4·30

The brickwork of a leading length is shewn in plate 7; wherein fig. 1 is a longitudinal section; and fig. 2 a transverse section taken on the line, A B, fig. 1—each representing the work as it appears as soon as the arch is keyed-in, and ready for the miners to commence excavating for another length onwards. In each of the figures the crown bars are shewn at F, the packings in brickwork at c, the props at G and H. B′ is the arch of the length last completed, and A′ is the arch of the preceding length; the line of junction between the two lengths being where the ends of the two sets of laggins meet, as at f, g. a′, b′, c′, are the three centre ribs, supporting the laggins upon which the last length was turned; and d′ e′, shew the leading and middle ribs under the laggins of the preceding length, the third or back centre rib having been moved onward to the position c′. The two ribs d′ e′ were left to remain undisturbed in the position shewn in the engraving, until the side walls of an additional length were built, when they with the lagging, sills, props, &c., were moved onwards, as named at page 125. D and D′ are the raking props to the upper and lower sills. The wedges or slack blocks, the half timbers, and the props, (all of which belong to the centreing), will be referred to and explained in Chapter XIV., which will be devoted to the particulars of the centres. Figs. 3 and 4 have already been referred to at page 114; where the brickwork is generally described, under the head of the side lengths.

THE JUNCTION LENGTHS.

In the manner now described the tunnel works are usually carried on in lengths, as they are called, of 9, 10, or 12 feet; as the nature of the ground will admit of. Twelve feet is a convenient length, in all cases where it can be adopted with safety; and this is done both ways from each shaft, till the workings meet; the last length, or space required to join the two workings, is called the junction (or thirling). The length of the junctions should be brought as near to that of the ordinary lengths as possible, so that the same timbers may be used; for if they be much longer, the same bars will not reach across from brickwork to brickwork, to take a bearing for the support of the earth; and if they be much shorter, there will be a difficulty in drawing the last crown bars from the last length each way: and, what is of more consequence, if the earth left between the workings which form the junction is but a mere thin partition, it will probably give way before the last two leading lengths are turned; and this might be productive of at least unpleasant consequences. It is the safer practice when approaching the junction to stop one of the workings, and advance to it in one direction only; for if the work be carried on at both sides, there is great probability of disturbing so narrow a wall of earth as would then be left for the junction length.

The only timbers required for a junction length are bars, and poling boards. The bars rest on the brickwork of the preceding lengths, and are built in, as the work advances. The side walls and the arch are constructed in the usual manner, together with the keying-in of the crown, as described at page 114, by the bricklayers inserting one cross lagging at a time, and closing each space over the said lagging, till at last he reduces the space to such small dimensions that he no longer can stand with his head and shoulders in it, to do the work,—as shewn in the engraving on the next page; he is therefore obliged in this last small piece to turn the top ring of the arch first, by fitting, and wedging his bricks tight, in the best manner that he is able, passing them with his hand and arm up the opening, and bonding the top into the next lower ring, by some of the bricks put up as headers; then setting the next lower course, or ring, bonding as before, and wedging it up tight; and so on with each course, until the opening in the bottom course, or soffit of the arch, will only be sufficient to receive one brick put endways into it; which brick, if necessary, must be tightly wedged into its place, with wooden or iron wedges, and the work will be finished. The whole of this final closing the work should be done with cement of the best quality. The space or opening left for this process need not exceed the dimensions of two cross or keying-in laggins. The annexed engraving represents a bricklayer, in the act of passing a brick up into its place, through a hole left by the omission of two cross laggins, in the manner above described. The closing portion, if properly done, is never likely to fall out, because its sides are splayed from the opening upwards, each course being wider than the lower one; they are also bonded into each other, as well as into the  toothings of the arch at each end of the aperture, and the last brick is wedged or rammed tight into its place so as not easily to be dislodged, added to which there is the adhesion of the cement, which, if good, and properly gauged, possesses great strength.

U

CHAPTER XIII.

TUNNEL ENTRANCES.

IT is unnecessary to state much upon the subject of the entrances; as it would answer but little purpose except to swell the volume. The designs for such constructions should be massive, to be suitable as approaches to works presenting the appearance of gloom, solidity, and strength. A light and highly decorated structure, however elegant, and well adapted for other purposes, would be very unsuitable in such a situation: it is plainness combined with boldness, and massiveness without heaviness, that in a tunnel entrance constitutes elegance; and, at the same time, is the most economical. The above conditions may be answered without cramping the taste of the Engineer, so far as taste enters into the composition of such designs; for architectural display, in such works, would be as much misplaced as the massiveness of engineering works would be, if applied to the elegant and tastefully designed structures of the Architect. Upon the London and Birmingham, and upon the Great Western railways, there are several very suitable structures of this kind.

The engraving on the following page represents the eastern entrance to Blechingley Tunnel; where the slopes of the open cutting are uniformly 2 to 1. The western entrance is similar in design; but the slopes of the open excavation differ from the slopes at the east end,—being $1\frac{1}{2}$ to 1 on the north side, and (by means of level benchings, is) 2 to 1 on the south side; which arrangement of the slopes was occasioned by the tendency of the strata to slip on the one side of the cutting more than the other, and arises from its dipping from south to north, and from the general disturbance of the beds, as explained at page 7. The quantity of brickwork in the wing walls of the one entrance is, consequently, greater than in the other: they are, however, similar in every other respect.

An inspection of the engraving will render it unnecessary to go into a minute description of the entrance, as the design is fully shewn in all its parts. The whole is composed of bricks; no stone whatever being used in the construction. Not but the use of stone for the string course, and the coping, would have been preferable to bricks for those purposes; but no good material of that kind could be obtained in the neighbourhood, and its cost when conveyed from London (about 24 miles, by land carriage, over a hilly country,) the nearest place from whence good stone could be procured, would

probably have been greater than was due to the difference in the comparative quality of stone and bricks, for that purpose.

The plan of the wing walls is circular, being struck with a radius of thirty feet, and battered three inches to one foot. The pilasters are six feet, and the plinth of the pilasters six feet six inches wide. The space between the pilasters, for the entrance of the tunnel, is 28 feet; and as the tunnel is 24 feet wide, the arch shows on the outside a thickness of 2 feet, and is formed of five half-brick rings. The space between the external contour of the finished arch, at the crown, and the underside of the string course, is one foot; the string consists of four courses of brickwork; and the height of the parapet, from the string to the under side of the coping, is three feet. The thickness of the coping is one foot three inches. A brick open drain, along the back of the parapet, conducts the rain water to similar drains made down the slopes, and hence to the water channels alongside the railway, where it joins the drainage from the culvert within the tunnel, (to be presently described), and passes off to the natural drainage of the country, at the tailing out of the open cutting.

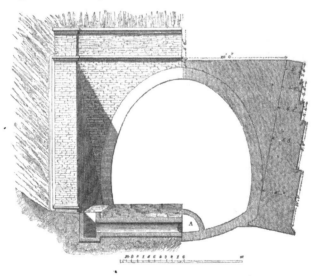

The Saltwood entrance differs materially from that constructed at Blechingley, and is shewn in the above engraving. The left-hand half of the figure represents the elevation of half the entrance; and the right-hand half shews a section of the same.

The wing walls are not curvilinear, as at Blechingley, but are built straight, or parallel to the line of railway, and follow the slope of the excavation; they also batter at the rate of two and a half inches to one foot. The dimensions of the wing walls, parapets, &c. are given in the engraving on the page opposite.

The following engraving is a section taken through the centre of the end of the tunnel; and shews a side elevation of the wing wall; and also a section of the barrel drain which conducts the water from the culvert that is constructed on the invert of the tunnel, throughout its length, similar to the one at Blechingley. A drain is also made along the back of the parapet and wing walls, to conduct the surface water to the proper drainage. By comparing the various parts of the preceding engraving with the corresponding parts of the following engraving, the details of the construction will be clearly apparent.

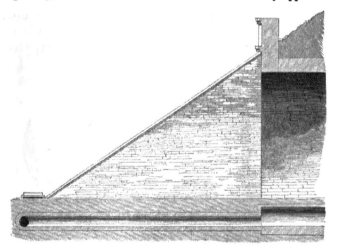

The erection of the entrances was paid for by the rod, of 306 cubic feet. The price per rod is not the same in all places; chiefly arising from the variation in the price of bricks. The quantity of brickwork in such entrances and wing walls as at Saltwood,—where the slope of the earth is 1½ to 1—amounts to 40 rods; and where the slope is 2 to 1, it will amount to 51 rods. This is reckoning for the work as shewn in the preceding engraving; but where there is additional work, as counterforts to the walls, &c. it will of course exceed this. The quantity of brickwork in the Blechingley entrance amounted to 65 rods,—the slope being 2 to 1,—and the wing walls being much larger than those at Saltwood, in all their dimensions.

THE SHAFT TOWERS.

At Blechingley Tunnel all the shafts (with the exception of No. 11) were left open for the purposes of ventilation; and at Salt-wood five were left open for the same purpose; the others were closed by doming them just above the arch of the tunnel, and filling them with earth to the surface. For the prevention of accidents, the brickwork of the shafts was carried up to some height above the surface of the ground, in the form of a tower, as shewn in the annexed engraving, and then covered in with an iron grating; which

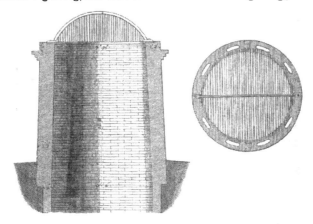

prevents stones falling down, if thrown for that purpose by mischievous persons. The domed shape of the grating not only gives it strength, but would cause such stones to roll off again.

The towers at Blechingley and Saltwood were raised 12 feet above the level of the spoil-bank, or where the brickwork of the shaft terminated. The shafts were of 9-inch work, but the towers were 14 inches thick; therefore, where the towers commenced they sailed over, outwards, 4½ inches, as shewn in the preceding engraving. The towers diminish upwards, from 9 feet at bottom to 7 feet 10 inches, inside dimensions, at the top.

The iron grating was made with a cast-iron circular ring, or plate, 9 inches wide, and an inch in thickness; with a lip around the inner edge that fits into the shaft, and keeps the grating from sliding from its bed. The plate was cast in halves, dovetailed together, when set in its place, as shewn in the right-hand figure of the last engraving.

An arched rib passed over the grating, which was bolted to the plate, and being also in two parts, was united in the middle by a half-lap joint, and a bolt and nut. The wires forming the grating were five-eighths diameter, and were riveted on each side, at the under side of the plate, and in the centre passed through the arched rib before named, which strengthened them, and kept them equi-distant apart. The cost of each grating, complete and delivered, was £6 10s.

THE CULVERT.

Along the centre of the invert of the tunnel, a culvert was constructed to carry off the water, both at Blechingley and Saltwood; its form and position is shewn at A in the engraving page 148 of the section of half the entrance of Saltwood Tunnel. It was described with a radius of 2 feet 3 inches from the centre of the invert, and was therefore not quite a semicircle, because of the rising of the invert on each side of its centre. The brickwork was 9 inches in thickness; and at Blechingley properly radiating or wedge-shaped bricks were made for it,—they were 10½ inches long, 4 inches thick at the back, and 3 inches at the inner side, and 9 inches in breadth.

To form the arch it required 26 of these bricks; and 90 were used in every yard forward, so that for the tunnel (1,324 yards long) 119,160 were used in the construction of the culvert. These bricks cost 50s. per thousand.

The sides of the bricks were bedded in mortar, but their ends were set quite dry, which left sufficient space for the water to drain from the ballast to the culvert. For Saltwood Tunnel it had been the intention to have constructed bricks of a similar kind, but the works having been let to contractors, they used common bricks for the purpose, of which there were required 264 in every yard forward. The bottom courses both of the culvert bricks at Blechingley, and the common bricks at Saltwood, were set in cement.

In order to facilitate the work, a centre was contrived 24 feet long for turning the culvert. This centre was fitted to a strong horizontal frame of the same length, and 3 feet 3 inches wide; and the whole was fitted to a similar frame, or stage, moveable upon rollers. The annexed engraving shows this culvert centre. Fig. 1, is a longitudinal section shewing six feet of its leading end, and fig. 2 is the elevation of the leading end. The laggins are shewn at A, which

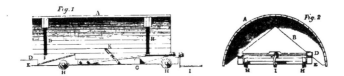

were made of inch deal. The centre ribs are shewn at B, B, and were six in number, made of one and a half inch stuff, and strengthened at the top by cross pieces, C. The upper frame is shewn at D, and E is the lower frame: the upper one is moveable upon the lower one, by sliding up the inclined plane, F, of which there are six on each side. When the upper frame is lying close upon the lower one, these inclined planes or notches resemble scarf joints. The frames are made of fir, 3 inches square, and are strengthened by cross ties, G, and diagonal braces.

The lower frame is moveable on twelve rollers, H, &c., (six on each side,) by which means it can easily be drawn forward, as the work advances. ɪ is a winch, the turning of which gives motion to a screw that is attached to the upper frame, while its thread works in a socket attached to the lower frame; when the screw is turned, it draws the upper frame forward, or causes it to slide lengthways upon the lower frame; but in consequence of its being notched, as it were, into the lower frame, in the manner of a saw-tooth rack, it must necessarily slide up these inclined planes; thus partaking of a compound motion, rising upwards as it advances forward; and when a reverse motion is given to the screw by turning the winch, ɪ, the contrary way, the upper frame, with the centre, will recede to its former position, and in so doing it will descend the inclined planes, and hence take a lower level. ᴋ shews one of six iron guides that partake of the motion of the centre, and keep it from moving sideways while it is being wound up, or the reverse.

By this contrivance the business of setting and easing the centre was greatly facilitated, and withal it was made twice the length that culvert centres usually are, thereby enabling the workmen to get on faster. In using it, the centre was first placed in position; the winch ɪ was then turned, till the segment was raised to the required level; and upon this the arch was constructed, which completed a length of 24 feet: now, by reversing the action of the winch, the centre was lowered from under the brickwork, and was therefore released from its pressure; whereupon it was drawn forwards by two men (nearly, but not quite), from under the 24 feet of culvert already completed; and being placed in line, motion was again given to the winch until the centre was raised once more to the proper level, which could now be known by the end of the centre fitting up to and under the completed length; a second portion of the culvert was then turned, and the centre raised and moved forward, as before. In this manner the work proceeded rapidly, as the easing, removing and resetting the centre did not occupy more than two minutes; and when the men got accustomed to it, they did it in but little more than a minute.

The following table shews the amount of labour, and the cost of constructing the culvert through Blechingley Tunnel—1,324 yards lineal—with the radiating bricks, and the above-described centre.

w

	Horses.	Bricklayers.	Labourers.
	Days.	Days.	Days.
Carting bricks	71	...	...
Lowering bricks and mortar ...	43½	...	68½
Turning culvert	..	151¾	167½
Loading brick carts, &c. ...	...	...	35
Wheeling bricks at top	...	...	252½
" in tunnel ...	...	...	44
Making mortar	...	...	30¾
Banking	...	...	16
TOTAL ...	114½	151¾	614¼

ACTUAL COST.

				£	s.	d.
	120,000 Bricks, including waste ...	at 50s. per thousand	=	300	0	3
Materials consumed.	75 bushels of Cement ...	at 1s. 8d. per bushel	=	6	5	0
	48 yards of Lime ...	at 13s. per yard	=	31	4	0
	96 yards of Sand ...	at 1s. per yard	=	4	16	0
	Carting ditto ...	at 2s. per yard	=	9	12	0
	Carting Water	... ...		2	5	6
	Candles ... 410lbs.	at 6½d. ...	=	11	2	1
Labour.	Horses ... 114½ days,	at 7s.	=	40	1	6
	Bricklayers 151¾ days,	at 6s. ...	=	45	10	6
	Labourers ... 614¼ days,	at 3s. 3d.	=	99	16	4
				550	12	11
	Centre, as above described			10	0	0
	TOTAL			£560	12	11

Being at the rate of 8s. 5½d. per yard forward.

The weight of iron work to the centres, including axles to the rollers, was 181 pounds.

Had the culvert at Blechingley been built in two rings, with common bricks, it would have taken 264 to a yard forward, or 350,000 in the whole.

When the tunnel at Blechingley was completed, it was cleared out from end to end, and the invert examined and made good where the raker-holes and props had injured it. A scaffold was made to run upon wheels along the tunnel, by means of which all parts of the arch and side walls were carefully examined, and its shape tested by a mould. Afterwards, the whole was limewashed, with a view to add to the light of the tunnel; which, as before stated, was but of little service. The ballast for both tunnels was mostly the *debris* from the brickyards, and the broken stone dug up from the temporary roads made on the works above. These were thrown down the shafts, and spread at the bottom. On the top of this rough ballast, a coat of finer material was used, about a foot in thickness, to bed the sleepers in; and the permanent way was then finally laid.

By means of the culvert above described, the whole of the water that might find its way into the tunnel was effectually carried off. The rough ballast that formed the bottom stratum was sufficiently porous to allow the water to pass to the culvert, which being built with the end joints of the bricks dry, admitted of the water percolating into the barrel of it, from whence it passed off to the drainage at the open end of the tunnel.

Towards the west end of the tunnel a considerable quantity of water entered through the roof at one place, and appeared likely to continue to do so:—it was, therefore, necessary to provide for conducting it to the under drainage, instead of allowing it to drop into the tunnel. This was accomplished by cutting a chase in the brickwork of the side walls, and, from the top of the said chase making a similar cut in the soffit of the arch to where the water was found to enter; then, concealing the chase or gutter, by bedding flat tiles in the front, to make the walls and arch of the tunnel appear perfect; this left a conduit or channel behind the tiles, to collect the water that entered the tunnel, and conduct it down the sides, to the invert, and from thence to the culvert, by a small drain. The tiles were set in cement, and the whole of the face was then plastered with the same material.

When the water enters at more than one place, but at no great distances apart, it can all be gathered into one general drain, by cutting oblique chases to collect, and lead it to one or more outlets.

At the Martello Tunnel, near Folkestone, where a great quantity of water enters through the roof, it is conducted to the sides by lining the roof with sheets of corrugated zinc.

CHAPTER XIV.

THE CENTRES OF ORDINARY CONSTRUCTION.—AND FRAZER'S PATENT CENTRES.

In the preceding chapters all particulars of the centres were omitted when describing the brickwork: an important and interesting part of the subject : the reason for so doing was to prevent confusion by introducing so many subjects together ; and as the account of the centres, and the method of using them, would occupy considerable space, it appeared better, for the general clearness of the subject, that it should be reserved for a subsequent chapter ; it being sufficient for the purpose of describing the brickwork, to state that the centres were set up, and the arch turned upon them. The particulars of the centres will now be given.

The centre ribs, that were made for and used in the construction of Blechingley Tunnel, were essentially the same as is commonly used in such works. It is frequently the case that contractors carry on their work with one set of centre ribs only, which must be taken from under the green brickwork to be set forward each time the arch of a new length is to be turned. As this mode of proceeding appeared objectionable, two sets both of centres and laggins were used upon the works under consideration ; by which means all danger and injury thereto was prevented ; and although the additional cost of such set of materials was considerable, yet it probably saved a much larger sum that would have been incurred in repairing broken work ; and ensured a sound tunnel.

When the Blechingley Tunnel was completed, the plant and other materials were conveyed to Saltwood, for the purpose of being again used, and the centres were partially so : but upon that tunnel being let by contract, and the contractors having appointed Mr. Joseph Frazer as their representative, that gentleman introduced his patent centres; thus affording ample means for observing and estimating the comparative merits of his centres, and those of the ordinary construction; especially as they were being worked under the superintendence of Mr. Frazer himself. Particular attention was, therefore, paid to the subject, that the result might be satisfactory. A description of each system of centres, with the cost of their construction, will be given; and a comparison drawn between them.

THE BLECHINGLEY CENTRES.—PLATE 9.

For each shaft of the tunnel there were required 10 centre ribs, 4 sets of laggins, 6 centre sills, 16 half-timbers, 40 props, and 40 pairs of slack blocks, or wedges; to these may be added a few wedges, and chocks, or chogs, &c.; the timber for which is generally obtained from the offal timber, which always collects in abundance on such large works, and therefore is not necessary to name in an estimate. The above quantity of materials is required for working from each shaft in both directions; consequently one half the above materials form a complete set, as used at Blechingley, for working in one direction, or as it is called "one end of the shaft." This will be described, as being all that is essential to the present purpose. It must be supposed that the side and shaft lengths are completed, and that the centres are to be applied to construct the arch of the leading work.

At each end of a shaft, five centre ribs are required; two of them must have no tie beam, as that would interfere with the raking props;—these were called segment or leading centres;—their construction and dimensions are shewn at large, fig. 3, plate 9: they consist of two segments, and when put together for use, are joined along the line, a, b, and also by the moveable tie, c; this tie prevents the spreading or contraction of the segments. This form of centre was found to be particularly convenient and strong, and as the only doubt about its utility, for all tunnel purposes, might appear to be the want of a complete tie beam at the bottom, this desideratum, if such it is, might be supplied by a moveable piece, or iron screw tie, to connect the point, d, to the opposite point, d'; if this were applied it probably would, in addition to the convenience of shifting and resetting, sustain any amount of pressure ever likely to occur, either vertically or laterally, and also all ordinary wear and tear, and damage from the blasting of rock, and therefore would require little or no repairs throughout a job. There is, however, an objection to moveable pieces, as they are apt to be mislaid and lost; but to prevent this, the men employed in the gang for shifting the centres should each be fined when a piece is lost, not merely to pay the value of the material, but also loss of time sustained by the brick-layers and others, in consequence of such neglect. At Blechingley the strength of these ribs was fully tested; for as they were the leading ribs, they were exposed to the greatest effects of the blasting, yet they never required any repairs; and after removal to Saltwood, by land and water carriage, they were uninjured; it would therefore appear that they are preferable to most of the ordinary centreing.

The other three ribs were differently constructed, and were called scarf or queen-post centres. Fig. 4 represents them: in construction they are well calculated to sustain heavy weights. This form of centre has been before used in tunnels, as on the Birmingham and Brighton lines of railway. The tie beam is a great security against the spreading or contracting of its span; but it is liable to interfere with the raking props, supporting the face of the excavation. Another objection to this form of centre is, that it is not well adapted to withstand the side blows to which it is exposed (particularly if used as a leading centre) when the miners are blasting rock or other material in the forward length. The extent of the repairs at Blechingley was considerable, not only arising from this cause, but being taken so completely to pieces each time they were shifted, they were more liable to injury, particularly in the scarf joint; whereas the segment centre is not so completely taken to pieces, for each of its two segments will pass through an opening of 5 feet 6 inches in width.

SETTING AND SHIFTING THE CENTRES.

As the bricklayers bring up the side walls, they leave holes about 13 inches deep (*a*, figs. 1 & 2, plate 9) at the proper places, for the reception of the ends of the sills; *a* in fig. 2 shews the hole from whence a sill has been removed forward. When the walls are at their full height, the sills, which are in two parts scarfed together in the middle, as shewn at A, fig. 1, are set in their places, and joined together at the scarf; the plates are then bolted, and the glands secured. Fig. 1 shews the sills, A, stretching across the tunnel, carrying the centres; and they likewise are shewn in section at A A A, fig. 2. Each sill has two props, B B, under it, the bottoms of the props are wedged on the invert, and their tops fit into a collar spiked to the under side of the sill as shewn at fig. 5, where A is the sill, B the prop, and *e* the collar; the first and third figures are an end and side view, and the middle figure is a plan of the under side of the sill, shewing the prop, B, nearly surrounded with the collar, *e*. The half timbers, C C, &c. were next laid on the sills nearly at right angles thereto: by comparing figs. 1 and 2, their situation and arrangement under the centres will be obvious. Each half timber was propped in the middle, or immediately under where the centre ribs take their bearing upon the half timber, as shewn at D D, &c. Instead of sills to support the centre, sometimes tressels are used; but at Blechingley the sills were adopted, as the most preferable mode.

The centres were next put together across the half timbers, and when set up in their required places, were raised to the proper level by the slack blocks or wedges, E E, &c., and also at large in fig. 6, where the wedges are shewn both in a side and end view. Upon the centres the laggings were laid, and thereon the arch turned.

It may be well here to call attention to the oblique manner in which the half timbers are placed, as represented in the engraving; one end of each of the middle half timbers, and the slack blocks, are under the heel of the leading rib, as at *b b* (fig. 1); whilst the other end of the same half timbers is under the queen post of the

scarf centre, as at *c c;* and as the queen posts in that kind of centre
are nearer the centre of the tunnel than the heel of the segment, or
leading centre, such oblique positions of the half timbers is necessary
to take those important bearings of the two ribs.

The setting of the three centre ribs, as above described, is all
that is required for each of the side lengths; where they ought to
remain after the arch is turned, during the time that the shaft length
is completed, and also the first leading length each way; when the
one nearest the shaft will be taken forwards without disturbing the
other two. By referring to plate 4, fig. 1—where the side lengths
are shewn as complete, and the excavation for the shaft length ready
for the brickwork,—it will be observed that each side length has all
its three centre ribs remaining under it; and by referring to plate 5,
fig. 1, the same shaft length is represented as complete: in this case
it will be seen that each of the side lengths still has its three centre
ribs undisturbed, and that the arch of the shaft length has been
turned upon four additional centre ribs, two of which will be carried
each way for the leading lengths, to make up the five centres required
on each side of the shaft, as described at page 124.

Upon the completion of the shaft length, the centres and laggings
were removed from under it, and a leading length on one side commenced; and as soon as the side walls were up springing high, the
centres were set without disturbing those in the side length, except
the back one of all (nearest the shaft) which was brought forward.
Fig. 1, plate 9, is a cross section of the tunnel, showing the work in
this state, and figure 2 is a longitudinal section of the same; the
cross section being taken near the face of the excavation through the
line, F G, looking towards the completed portion of the tunnel.
The brickwork is shewn as being above springing high, and part of
the arch turned, as at H H, six of the laggins on each side having
been already laid on the centres.

Figure 2 shews the five centre ribs in use at one time; three
under the length in progress, and two supporting the laggins under
the last turned length; the third or back rib having been moved
forward for use under the advanced work. When the arch is
completed the whole five ribs remain unmoved, under the two lengths,

(as shewn in fig. 2,) during the whole of the time another length forward is excavated; and also until the invert and side walls of the same are constructed; thereby materially assisting the new brickwork to sustain whatever pressure it may have to bear, not only from the earth above itself, but one half the weight of the newly excavated length; as may be seen in fig. 2, where the brickwork, κ, has to carry half the weight of the new length, because one end of all the bars rest upon it near its end, as at L. Furthermore the invert and side walls being constructed before any of the centre ribs are removed, they form a buttress against the completed work, that prevents any tendency in those lengths to slide forward and separate from the preceding work; for it sometimes happens that nearly a whole length will move onward, and leave an open joint between itself and the length it had separated from; for in all these operations the work has a tendency, as before stated, to press forward in the direction that it is being carried on. None of the support of the arch of the two preceding lengths were ever disturbed at Blechingley Tunnel, till wanted for a third length; for although the back rib of all was moved forward as in fig. 2, yet the laggins of the said back length were kept tight up to the arch by the remaining two ribs, so that two lengths of 12 feet each, or 24 feet of completed work, remained with its full support, not only till the next length was excavated, but until the next side walls were up ready for the centres.

Under these conditions every length was well able to bear the superincumbent weight, until it received assistance from the neighbouring advancing length; the construction of which to the springing height necessarily occupied some days, and therefore the cement had time to harden before the weight came upon the arch after the removal of the centre ribs; which is an important advantage.

But when, from motives of economy, three ribs and one set of laggins only are used, the whole support must be removed from under the first length before a second one can be turned, and again must be in like manner removed from the second before the third can be constructed, leaving the back work without support; which occasionally causes it to give way,—the bricks to crush,—and frequently the two last lengths to be separated from each other;

x

and if the arch does not come in, it is often bulged in various directions, and consequently unsightly if not unsafe. Enough has now been stated to show that, in heavy ground at least, the cheapest method of proceeding is not the best, or, at all events, not the safest.

The cost of setting the centres and removing them forward was included in the price of the brickwork, namely £4 10s. per rod. This work was performed by a gang of men who devoted their whole time to it, and contracted with the several bricklayers to do this part of their work: and thus by constant practice they obtained a readiness in its execution.

The price paid by the bricklayers at first was £3 per set, but this was subsequently reduced to £2 10s.—the centre setters finding all necessary tools, candles, &c.

The following will show the total cost of a double set of centres, and the requisite materials for one end of a shaft, the work proceeding as at Blechingley Tunnel.

		£	s.	d.
2 leading or segment centres ... at £10 12s. 11d. each		21	5	10
3 scarf ditto ... ... £9 4s. 6d.		27	13	6
2 sets of laggins ... ... £6 10s. 0d.		13	0	0
2 sets of keying ditto ... ... £1 0s. 0d.		2	0	0
3 centre sills, and ironwork complete		13	5	6
8 half timbers		4	7	11
14 props		3	19	4
6 collars		0	6	0
40 slack blocks		2	0	0
Total Cost of Materials		£87	18	1

The preceding prices were paid at Blechingley, in 1841. But subsequently the duty on foreign timber has been greatly reduced, and therefore the same materials at the present time, 1844, would not cost so much by a considerable amount.

FRAZER'S PATENT CENTRES.—PLATES 10 AND 11.

A set of Mr. Frazer's centres consists of three ribs, which are represented as in use, in plates 10 and 11. In fig. 1, plate 10, which is a transverse section of Saltwood Tunnel, they appear, as viewed from the face of an advanced or leading length, when such length is completely excavated, and timbered ready for the bricklayers to commence the construction of the invert. The centres are in the position they held when the arch of the last length of brickwork was turned, and therefore they appear to be supporting the said arch. Fig. 2 is a longitudinal section of a portion of the tunnel, shewing the same state of the work, and consequently at a time when the centres are at rest, waiting to be advanced onward as soon as the invert of the next length, and the side walls thereof, up to the height of the springing of the arch, are constructed. Fig. 1, plate 11, is also a longitudinal section, shewing the invert, and side walls; and the centres advanced and adjusted to their places, ready for the bricklayers to commence upon the arch.

The three ribs are distinguished in the following description by the letters A, B, and C; A being the leading rib, B the middle rib, and C the back rib. Each rib is constructed of elm timber, $4\frac{1}{2}$ inches in thickness, and 16 inches wide; and consists of four pieces scarfed together, as shewn in figures 1 and 4, plate 10. In the ordinary construction of centres, the ribs when the laggins are upon them are all of one size, and of the same span and rise as the under side (or soffit) of the intended arch; but in Mr. Frazer's centres all the three ribs differ from each other in dimensions of their radii, and the middle rib is the only one that acts in the same way as centres of ordinary character, namely, having the laggins and the arch immediately resting upon the rib, and consequently, with the laggins, is of the same dimensions as the arch (in the clear). The leading rib is larger, and the back rib smaller, than the said dimensions. Each rib will now be described separately.

The leading rib, A, is $12\frac{1}{2}$ inches larger radius to its outer edge than the under side of the arch, and $3\frac{1}{2}$ inches less radius to its inner edge; both edges are covered with half-inch iron plate, bolted quite through the rib, see fig. 4, where the plates are shewn at $a\,a$, and the bolts at $b\,b$, &c; the plate on the under side is 6 inches wide, and is placed to project 2 inches over one side of the wood, forming a flange for the laggins (which are 3 inches thick) to rest upon, as shewn in section, figs. 2 and 5; the former, being a longitudinal section taken along the centre of the tunnel, shews the *keying-in* laggins, resting upon the flange, and the latter shews the *long* laggins so resting, wherein A is the rib, *d* the iron plate projecting over the rib, inwards, upon which projection the laggins, *e*, rest.

When the laggins are in their places their upper surface forms the core or bed upon which the arch is to be constructed. The leading rib, when set, must be its whole thickness in advance of the end of the intended length of brickwork, and therefore it will stand in front of, or cover 12½ inches of the toothing end of the same; this forms a convenient mould, to guide the toothings of the work as they are brought up, the same being set out, and chocks of wood being nailed thereon, compels the bricklayers to keep their work regular. The chocks are shewn at *c c c*, &c., fig. 4, which figure shews the inner face of the rib.

The ribs, B and C, rest upon and are fixed to a trestle, D, on each side of the tunnel; but it will be seen, figs. 1 and 4, plate 10, and fig. 2, plate 11, that the leading rib, A, is supported by wedges or slack blocks, *d*, upon the end of the brickwork of the side walls, E, which for this purpose is carried up nine inches longer than the arch of the said length is intended to be. This, Mr. Frazer states, was all the support that he ever found necessary to carry the said rib and its superincumbent weight; but at Saltwood Tunnel, where the ground was heavy, and the bricks none of the best quality, the weight pressing upon the said rib, occasionally broke off the end of the brickwork upon which it rested; in cases, however, where all circumstances are favourable, the plan of resting the rib upon the end of the brickwork may perhaps be sufficient, but it would certainly appear to be by no means a safe method of proceeding without the support of the props, F, which will next be described.

This prop is represented at F, figs. 1 and 2, plate 10, and at large figs. 2 and 3, plate 11; the upper part of the two latter figures shews two views of the top of the said prop supporting the rib, A, and the lower part shews the manner in which the prop is supported upon the invert, part of the skewback of the invert being cut away to receive an iron block, G, for the square end of a capstan-headed screw-bolt, *f*, to rest upon, the screw of the said bolt being the means of tightening up the prop when set in its place under the rib; the foot of the rib is held between two iron cheeks, *g g*, which are fastened to the top of the prop by a bolt, *h*, and a collar, *i*, figs. 2 and 3, plate 11. The end of the upper slack block passes between the iron cheeks, and appears at *k*, figs. 2 and 3, thus affording facilities for striking the same when the rib is to be eased; the under slack block does not extend within the iron cheeks, but is cut off at about the level of the brick-work, as at *l*, fig. 2.

Before the prop can be applied to the rib to assist the brickwork in carrying it, the rib must first be set and tightened up, or adjusted in its place, ready for the reception of the laggins and the brickwork; such adjustment being effected by driving or easing the slack block or wedges, because the prop is in the way of any alteration of the said wedges, after it is fixed and tightened up to the heel of the rib, by the screw *f*, at the bottom of the prop.

The middle rib, B, (and also the back rib, c) stands upon, and are permanently fixed by brackets, and straps and bolts to the trestles, D, as shewn at figs. 6 and 7, plate 10, and moves forward with it upon the rollers, *m*, figs. 1, 2, plate 10, and 1, plate 11; the under side of the middle rib is covered with half-inch plate iron in one piece, bolted through, as shewn in fig. 7, plate 10, which gives strength to the arched rib, in the same manner as the struts, &c. apply in centres of the ordinary construction. The bolt and nut are shewn at large, fig. 9. The laggins, *e e*, &c. rest flat upon the upper edge of the middle rib, and therefore the radius of this rib must be the same as of the arch, all but three inches (the thickness of the intervening laggins,) and as before observed, it is the only rib of the three that carries its load like the centres of the common kind.

The back rib, c, is also strengthened with a covering of half-inch iron in one piece on its under side, bolted through like the middle rib, with this difference, that patent screws, fig. 12, are substituted for every alternate bolt, as they are cheaper than the bolts: between each bolt and screw, a hole is made quite through the rib, to receive the stem of a bearing iron, *n n*, &c. fig. 6, and at large, fig. 10; and there are as many bearing irons as there are to be laggins. It will be seen, fig. 6, that the laggins do not rest upon the back rib itself, but upon the bearing irons, which project from the timber, or upper edge of the rib; the amount of projection being regulated as may be required to press the laggins to their proper level, by means of adjusting screws, *o o*, &c., which are tapped into the half-inch iron plate, and act upon the stems of the bearing irons; and, conversely, the reversing of the said screws, lower the bearing irons and ease the laggins from under the brickwork, one by one, to be removed forward.

As before stated, the ribs, B and C, are permanently fixed at their footings to the trestles, D; and they are also steadied at their crown by the irons, H H', as shewn in fig. 2, plate 10, and fig. 1, plate 11. These irons are moveable at one end, which, forming a hook, drops into an eye screwed into the side of the rib; the bricklayer, by unhooking either of the irons can put them out of their way, when the work advances towards the crown of the arch; each iron however, can only be unhooked at one end, the other end being permanently fastened by the eye to the other rib, so that when unhooked by the workmen, they hang down as shewn at H', fig. 2, plate 10. If they were not so attached to the ribs they would very soon be mislaid and lost.

The trestles, D, with their load, move upon rollers, *m*, and half-timbers, I I, figs. 1 and 2, plate 10, and fig. 1, plate 11, laid longitudinally, as a kind of tramway for the rollers to run upon: the half-timbers are held in their places on the skewback of the invert, by bricks or blocks let therein for that purpose.

SETTING AND SHIFTING THE CENTRES.

In setting these centres for the turning of the arch, the leading rib must first be set in its place, and wedged up, on the end of the brickwork, E, until it is at the correct level; the prop, F, is then to be placed, and screwed up tight under the heel of the centre, which will readily be understood by those who have attended to the previous description; next, the trestles, D, are rolled forwards, until the ribs, B and C, are advanced to their proper places; when this is done, three pair of wedges and blocks, K K K, are placed between each trestle and the half-timbers, and by their use the trestles are lifted up, until the top of the middle rib is upon a level with the iron flange of the rib, A, thus forming two level bearings for the laggins; next, the bearing irons of the ribs, C, must be pressed outwards, by the adjusting screws, o o, until the top of each of them is also upon the same level; so that when the laggins are placed one by one upon the three ribs, they will bed solidly upon them all; the adjustment of the bearing irons or of the three ribs may be tested by passing a laggin over each of the irons and the other two ribs in succession, for the first length of the brickwork; but in all succeeding work their adjustment is regulated by the brickwork last completed, as will be shewn presently. The three bearings of each laggin, when the ribs are adjusted for the intended arch, is as follows :— 1st, *upon the iron flange of the leading rib*, 2ndly, *upon the middle rib itself*, and 3rdly *upon the bearing iron of the back rib;* when this is all made correct, the centres will be ready for the bricklayers to turn the arch, and by whom the laggins are laid on one by one as the work advances.

When the arch is completed the miners again take to the work, and excavate and timber another length, as shewn fig. 2, plate 10; as soon as this is done the bricklayers re-enter and construct the invert and side walls to the springing height; the centres have then to be advanced for turning the arch. Fig. 1, plate 11, shews this stage of the work; the side walls are represented as up,' and the centres in their places; four laggins are also shewn as drawn forward, and the work of the arch commenced. The detail of the method of moving forward the centres now remains to be given.

First a rib, called a jack rib, L, is fixed under the laggins, in the rear of the back rib, C, fig. 2, plate 10; this jack rib consists of an iron plate, 1 inch in thickness, and about $2\frac{1}{2}$ inches wide, which is bent into a form of the arch of the tunnel. Fig. 8, plate 10, shews a portion of the jack rib upon a larger scale, wherein it will be seen that, opposite to every alternate joint of the laggins, a screw is tapped into and passes through the rib; which screws have their

outer end finished as a loop, whereby they may be turned with a
lever; their inner ends are finished with a square head driven on to
the end of the screw, and are similar in appearance to the ends of
the bearing irons of the back rib, c, before described, but not so stout,
neither do they revolve, as the screw is turned round.

The screws being placed opposite every alternate joint of the
laggins, it is clear that only half the number of screws that there are
laggins, are required, as each screw presses against two of them at a
time, and exerts quite sufficient force to retain the laggins in their
places when the ribs, b and c, are removed from under them ; and
which is all that is required of the jack rib. This arrangement is
fully shewn at fig. 8. $r\ r$, &c., are the screws passing through the
iron plate, s, that forms the rib, the square end of each screw pressing
two laggins, $e\ e$, &c., at their joints, against the brickwork of the
arch, m. The jack rib terminates with a screw, o, which works in
sockets, as shewn in the figure, and its extremity rests upon an iron
bar, n, about two feet long and four inches wide, driven temporarily
into the brickwork, for the purpose of carrying the rib.

The jack rib will be understood to be only a flat iron arch,
springing from the iron support, n, on each side ; and its use is
simply to support the back ends of the laggins by means of its screws,
$r\ r$, &c., when the trestles, d, with the two ribs, b and c, are removed
from under them by lowering of the wedges, k. The other ends of
the laggins are, at the same time supported by the leading rib, a,
independently of the trestles, and therefore, under such circumstances,
the laggins will continue undisturbed. They would, however, no
longer be of use in preventing the arch above them from giving
way, if it should be subjected to extraordinary pressure before the
cement is well hardened ; the leading rib being the only part of the
centreing that would afford assistance in such a case. The use of
the screw, o, fig. 8, is for the adjustment of the jack rib to its required
elevation under the arch, or the laggins.

As soon as the jack rib is fixed, and its screws, $r\ r$, &c. adjusted,
to take the ends of the laggins or press them against the brickwork,
of the arch ; the iron bar H' is unhooked, and left to hang down, as
shewn at fig. 2, plate 10, and the wedges, k k, &c. are eased, whereby

the trestles are lowered, and their rollers, *m m*, rest upon the wooden tramway laid for their reception; which tramway being then continued in the direction that the work is proceeding, the trestles are rolled onwards; and with them, of course, the middle and back ribs, B and C, which are thus advanced to their proper position for the turning of a new length of the arch. When the ribs, B and C, are lowered as above described, the middle rib, B, easily passes under the leading rib, A, which would yet remain in its place, under the toothing end of the last-turned arch.

It will be remembered, that the leading rib, A, was described as placed in front of the brickwork of the arch, its iron plate carrying the ends of the laggins; this arrangement, which is shewn in section at figs. 2 and 5, plate 10, is contrived in order to keep as much clear headway as possible, for the passing of the middle rib under it, when that rib is advanced, with the trestles, to the next forward length.

The trestles, with the ribs, are moved forwards until the back rib, C, nearly abuts against the rear edge of the leading rib, A, or, more particularly, it is moved till it is within six inches of the ends of the laggins, when they are again wedged up to a proper level, as before; the bearing irons of the back rib, C, are then pressed, by their screws, home to the laggins of the arch, which afford them, and the superincumbent brickwork, the same support at that end that had hitherto been obtained from the leading rib, A; whereupon the said leading rib may be released, by easing the wedges, *d*, at its springing, and moved onwards by passing it over the rib, B; it is then to be fixed as before, on the end of the brickwork of the newly-constructed side walls, and adjusted, together with the rib, B, to their proper level, which may be proved by ascertaining that the laggins of the arch already completed, *if produced*, would fit on the ribs, A B, in the same straight line. The props, F, &c. must then be again fixed on each side, to support the heels of the rib, A.

When the centres are set, and adjusted to their proper level, as above described, they are ready to receive the brickwork of the arch; for which purpose the same set of laggins are again used, they being drawn forwards from their last place, one or two at a time, as they are wanted, beginning at the springing. One of the screws of the

Y

jack rib, and the corresponding screw or bearing iron of the back rib, c, are eased, which releases the laggins against which they pressed, and admits of their being drawn forward to their places on the advanced centres; the screw of the bearing iron of the back rib is then again tightened, which presses the rear end of the laggin tight against the arch, or under the toothing end of the last-turned length. The three bearings of each laggin, as it is so advanced and placed, will be as in the former case, (if the centres are correctly adjusted in position) *on the iron flange of the leading rib, A, on the middle rib itself, and on the bearing irons of the back rib.* Upon the laggins as they are thus drawn forwards, the bricklayers construct the arch.

From the account above given, it will be obvious that (in using these centres) so soon as the arch of one length is completed, it remains supported by its centres and laggins, until another length is prepared by the miners, and the invert and side walls thereof are built; whereupon the two centres, B and C, are removed from under the said arch, leaving its toothing end only supported; for the jack rib does nothing more than keep the laggins from falling down. The back rib, c, is next brought under the toothing end, and its bearing irons screwed up tight to the laggins, which enables the leading rib, A, to be released, and carried forwards; so that, from the time that the side walls of a second length are constructed, the last turned arch receives no support, except under its toothing ends.

The following table shews the total cost of a complete set of Frazer's patent centres and the necessary materials for one end of a shaft, as used at Saltwood Tunnel.

			£	s.	d.
Leading rib	... ...	...	9	3	3
Middle rib	... .	...	6	6	8
Back rib	... ...	...	9	11	3
Jack rib	... ...	...	7	16	8
Irons for ribs	... ...	...	0	4	0
2 props for leading ribs	...	£1 3s. 2d. each	2	6	4
2 trestles	... ...	£8 18s. 10d. each ...	17	17	8
1 set of laggins,—same as at Blechingley ...	...	6	10	0
1 set of keying-in ditto	ditto	... ...	1	0	0
	Total Cost of Materials	... ·	**£60**	**15**	**10**

COMPARISON.

A double set of Blechingley centres, with all the necessary materials to be used, as already described, cost as stated at page 162, the sum of £87 18s. 1d. If a single set be employed, as they might be with safety, where the ground is light, (and which is the plan frequently followed by contractors), the cost would be £50 6s. 11d.; which exceeds half of the above sum, because more than one half of that quantity of materials would be required.

The cost of a corresponding set of the patent centres would be, as shewn above, £60 15s. 10d. (the prices of the materials being taken alike in both estimates, for the sake of correct comparison). The patent centres would therefore be less costly than a double set of ordinary centres, by £27 2s. 3d., and more costly than a single set, by £10 8s. 11d.

It may also be considered, that, when the work is completed, and the centres laid aside, the sills, half-timbers, and props, used with the ordinary centres, would be worth more money, as timber, than the pieces forming the trestles, &c. of the patent centres; because they would be less cut up into small pieces. The former, indeed, would be nearly as valuable as when first put into use for those purposes, if proper care had been taken of them. The real deduction from their original value would be the usual charge for use and waste.

Where the ground is heavy, the use of a single set of materials of the ordinary construction, has been explained, in the preceding pages, as being an unsound mode of proceeding; but this is by no means the case when the ground is light; for then one set of centres will be sufficient.

The patent centres, when used in heavy ground, do not afford anything like the support to the brickwork that is derived from the use of a double set of those of the ordinary construction. This opinion would be arrived at by the perusal of the preceding pages, and is in accordance with the author's experience, and observation of their use, at Saltwood Tunnel.

For ground that is light the patent centres are well adapted; and even when the ground is moderately heavy they may be advantageously employed, for they afford more security to the work than a *single* set of the ordinary materials, although they fall much short in that particular of a *double* set.

The great advantages the patent centres appear to possess over centres of the ordinary construction, is in the total absence of queen posts, and struts, &c., (which form the framework of the last-named centres), they therefore leave a large open space for the scaffolding and materials of the bricklayers, who can get at their work with greater facility, by having so much more room than is afforded by the ordinary centres. Another advantage, and which in tunnelling through rock, where much blasting is necessary, is great,—namely, that the debris from the explosion is less likely to disturb or injure these centres; whereas, with the other kind of centres, it is likely to do much mischief, and occasion considerable outlay in repairs.

Independently of the difference in the first cost of the two kinds of centres, it does not appear to the author that any money is saved in the subsequent use of the patent centres; both kinds being equal in this respect.

Although it would appear, from the above statement, that a set of the patent centres, and necessary materials, are rather more expensive, in their first cost, than a single set of centres and materials of the ordinary construction, yet there can be no doubt that they are greatly superior thereto, in their practical application; for although one set of centres and materials will be sufficient for tunnel works in light ground, yet there is no ground so perfectly homogeneous in character throughout a hill, but in some parts, from faults, or local disturbance of the beds, will prove heavier than other portions of the same strata. When this case occurs, and the patent centres are in use, no cause for alarm need be apprehended, there being sufficient provision therein to withstand the effects of moderately heavy ground. The difference of the original cost ought therefore, not to be considered, as compared with the greater security obtained by the use of them.

It must, however, be understood that the author's strong recommendation of Mr. Frazer's centres is limited to *light* or *moderately heavy* ground; for where the earth is very heavy, the use of a double set of materials of the ordinary construction is greatly superior thereto. Under such circumstances, the extra cost in the original outlay would be nothing in comparison with the superior advantages to the security of the work during its construction. In short, those who have had experience in tunnelling in heavy ground would consider no expense (within reasonable limits) too great, to secure the work from such casualties as it is liable to under those cir-- cumstances.

CHAPTER XV.

CAST-IRON SHAFT CURBS.—CUTTING BRICKS.—WEIGHT OF BRICKWORK.
BRICKMAKING CONTRACT.—ETC.

WHEN describing the brick curbs that connect the shafts with the soffit of the arch of the tunnel, it was stated (at page 126) that the subject would be again returned to, in order to describe the curbs of cast iron. The reason why those particulars were not inserted in that part of the work, was on account of an unavoidable delay in the preparation of the engraving, and it was deemed unadvisable to delay the printing until that was done, as there would remain the present opportunity to supply the omission.

Curbs of cast iron were used by Mr. Stephenson, in the tunnels on the London and Birmingham Railway, and the engraving, plate 12, represents them in detail.

Fig. 1, is a plan of the curb, as it is fixed ready for the support of the shaft. It is made in four segments, which are fitted and bolted together at a, b, c, d. The clear diameter of the circular area, formed by the curb, is the same as that of the shafts, nine feet; and the top of the curb is a level surface (or flat ring) 1 foot 3 inches wide, and upon this the brickwork of the shaft rests. The dotted circular line shews the junction of the curb with the brickwork, as it appears from below, and corresponds with the point e in each of the other figures.

The brickwork of the shaft is built flush with the inner edge of the curb.

Fig. 2, is a section taken through the curb, upon the line, A B, of the plan, or at right angles to the direction of the tunnel. (This section corresponds with that of the brick curb, shewn at fig. 2, plate 5.) The under side of the figure represents the soffit of the arch, and the line, $e f$, is the skewback supported by the brickwork of the tunnel, E. The line $e f$ is, in this part of the curb, at its greatest obliquity, or makes its greatest angle with a vertical line; from thence its obliquity gradually diminishes (each way) throughout a quarter of a circle, or, to the crown of the arch in the direction of the tunnel, where the line, e f, becomes perpendicular.

Fig. 3, is a section on the line, c D, of the plan, and is at right angles to the section in fig. 2. The line, e f, in this figure, where it joins the brickwork of the arch, E, is, as above stated, perpendicular. (This section corresponds with that of the brick curb shewn at fig. 1, plate 5.)

Fig. 4, is a back view of the curb, looking at it in the direction A B, and shews the manner in which it is formed to abut against the courses of the brickwork of the arch.

Fig. 5, is also a similar back view, but taken in the direction c D, or at right angles to the last figures.

The curb is cast with chambers, in order to combine lightness with the requisite strength.

A plan of the brick curb is given at fig. 3, plate 5, and by comparison of the two sections in that plate, with the engraving, plate 12, together with what has been stated upon the subject, there can be no difficulty in comprehending the whole of its details.

CUTTING OF BRICKS.

In several parts of the preceding pages it has been stated, that when wedge-formed bricks were required for particular purposes, as for the skewback of the tunnel invert, they were either moulded to the required shape, or the common bricks were cut for that purpose; when the latter plan was adopted, they were cut in the following manner, which saved a deal of time as well as waste of the material. A number of them were placed in a box (which was purposely contrived) and then wedged, or screwed up tight; they were then cut with a stone-mason's saw, working through a saw kirf in the opposite sides of the box, at the required angle, in a similar manner that a joiner cuts the mitres of his mouldings. This method answered very well, and had been previously used by the author for cutting bricks to the proper angle for the face of oblique arches.

WEIGHT OF BRICKWORK.

An experiment was tried, on September 3rd, 1842, to determine the weight of a cubic yard of brickwork. On the works at Saltwood there was an excellent weighing-machine, by Pooley & Son, upon which the experiments were tried.

BRICKWORK IN CEMENT.

				Tons.	cwt.	qrs.	lbs.
A cubic yard of dry bricks ...	...	(384)	=	1	2	1	20
Sand, water, and cement for ditto		...	...	0	6	2	4
Total weight of a cubic yard of brickwork in cement			=	1	8	3	24

BRICKWORK IN MORTAR.

				Tons.	cwt.	qrs.	lbs.
Bricks ... as above	...	...	...	1	2	1	20
Mortar for ditto ...	...	...	...	0	4	1	8
Total weight of one cubic yard of brickwork in mortar			=	1	6	3	0

BRICKMAKING CONTRACT.

The following is a copy of an agreement made with one of the contractors for the brickmaking at Blechingley :—

Memorandum of an agreement made this 19th day of March, 1841, between William Chaplin, brickmaker, of the one part, and Frederick Walter Simms, on behalf of the South-Eastern Railway Company, of the other part. The said William Chaplin agrees to dig, make, burn, and deliver, as many bricks as the said Frederick Walter Simms, or other the Resident Engineer to the said Railway Company shall require, without intermission, at or near to the proposed Blechingley Tunnel, on the line of the said Railway, and upon such land or lands as the said Frederick Walter Simms may require, and to find at his own expense all labour, tools, utensils, horses, and every material except coals, that may be required in and for the production of kiln-burnt bricks, of the best quality, and also to build and maintain in repair all necessary kilns for the burning of the said bricks, he finding all labour and materials for that purpose, with the exception of bricks, which are to be found or provided by the said Company,— for and in consideration of the sum of fourteen shillings per thousand, for all bricks which he may deliver from the kilns; and the said Frederick Walter Simms agrees, on the part of the said South-Eastern Railway Company, to pay to the said William Chaplin the sum of fourteen shillings per thousand, for so many bricks delivered from the kilns as the said Frederick Walter Simms may require; the said South-Eastern Railway Company to find coals, and to pay the duty only, and to allow the said William Chaplin the use of such soft bricks as may be required to build the kilns and maintain them in repair, and to sink a well if required. And it is hereby further agreed, that no payment shall be made to or required by the said William Chaplin, on account of any work in progress, or bricks in the course of making; but that he shall be paid only for such bricks as shall have been previously delivered to the said Frederick Walter Simms, as above specified. And it is also further agreed, that the bricks shall be of the best possible quality, and of the largest dimensions allowed by Act of Parliament; and that if the said William Chaplin and the said Frederick Walter Simms shall disagree upon any point regarding the quality of the bricks, or the mode of carrying on the work, or upon any other matter whatsoever, the question in dispute shall be referred to William Cubitt, esquire, or other the Engineer-in-chief to the said South-Eastern Railway Company for the time being, whose decision shall in all cases be binding and conclusive; and that, if the said Engineer-in-chief should think proper to put an end to this agreement, and determine the connexion between the said William Chaplin and the said South-Eastern Railway Company, before that a sufficient number of bricks shall have been made or provided for the said proposed tunnel, whether made by the said William Chaplin or otherwise, he shall be at liberty to do so, upon the said Company taking the unfinished stock, and the materials that may be upon the ground, at a fair valuation, unless the cause of the said Engineer determining and ending of this agreement shall be or shall have

z

been occasioned by neglect on the part of the said William Chaplin to push forward the work entrusted to his care with the utmost possible expedition; or if he shall have executed any part or parts of the said work in an inefficient or unworkmanlike manner; or from any other reasonable cause,—then, and in such case, it shall be at the option of the Engineer whether or not the said Company shall take to the said materials. In case of a valuation of the materials becoming necessary, such valuation shall be made by two referees, one to be appointed by the the said William Chaplin and the other by the said Company, or by an umpire to be appointed by such two referees in case they should disagree; and in case of either of the parties neglecting to appoint a referee within seven days after being required in writing by the other of the said parties so to do, then at a valuation to be made by the referee of the other of the said parties alone, whose decision shall be binding and conclusive.

In witness, &c.

ESTIMATED COST OF HORSE-POWER EMPLOYED IN WORKING THE GINS

During the Shaft-sinking and Water-drawing at Saltwood Tunnel. 1842.

Expense of 67 horses, and attendants, in twenty-four hours.

			£	s.	d.
2 quarters 1 bushel of beans	...	38s. per quarter	4	0	9
2 quarters 1 bushel of oats	...	25s. „	2	13	1
50 trusses of hay ...	£5 10s. per ton of 40 trusses		6	17	6
40 trusses of straw ...	...	8d. per truss	1	6	8
Shoeing each horse ...	...	per diem 1d.	0	5	7
Farrier's expenses ...	per horse „	3d.	0	16	9
Stabling	per horse „	3d.	0	16	9
Harness and repairs ...	per horse „	3d.	0	16	9
12 stable men	... „	3s. each	1	16	0
18 gin-boys	... „	1s. 3d. each	1	2	6
18 ditto	per night, 1s. 6d. each		1	7	0
			£21	19	4

Being at the rate of 6s. 6⁷₁₀d. per diem for each horse.

They were supplied with as much food as they could eat, not only in the stable but at every interval of rest during the time of working.

The average time made by each horse was 1·11 shifts per diem, which is 7s. per shift, gave 7s. 9$\frac{1}{10}$d. as the earnings of each horse. Leaving 1s. 2$\frac{1}{2}$d. per diem to cover contingencies arising from the death or depreciation in the value of the cattle.

The above table will give a good approximation to the quantity of food consumed by horses when working hard, and of the general expences attending that kind of work. The cost of the corn and hay will of course vary from time to time. When the works were executed, in which the horses above referred to were employed, the price of grain, &c. was very high.

APPENDIX.

BLECHINGLEY TUNNEL SICK FUND.

WHERE a large body of men are collected together, as upon the works of a tunnel, there will be sure to arise considerable sickness among them, as well as occasional accidents. To provide for, and defray the expences of the necessary medical attendance, &c., it is the usual practice to raise a fund by means of a small contribution from the weekly earnings of every one employed upon the work; and therefore a set of rules are necessary to provide for its proper distribution. The following are the rules and regulations of the Sick Fund that was established at Blechingley Tunnel; which, upon the whole, were found to answer very well; but, in the course of time, two additional rules were found requisite, and were accordingly added: these are annexed below. Besides these, there appears to require an alteration in the third rule, which enacts that all members shall contribute alike, namely, "sixpence per week." This would have been better if it had required each man to pay a per centage on his earnings, as for instance, one farthing, or one half-penny in the shilling, according to the requirement of the sick list. Such an arrangement would have been more equitable than the one adopted.

There should also have been a small fine attached to the non-attendance of the Committee-men at the appointed time of meeting, as their want of regularity in this respect gave additional trouble to the Treasurer, and to the Clerk who kept the accounts of the fund.

Except what has now been stated, with the two additional rules before named, nothing more appeared wanting to make the regulations complete; and with a view to their being useful on any similar occasion, they are inserted in this volume.

RULES AND REGULATIONS.

1.—This Fund is to be formed by subscription of Workmen employed in the construction of Blechingley Tunnel, or in connexion therewith, for their temporary relief in sickness, and in case of accidents, and for the payment of Medical Assistance.

2.—A Committee of Management shall be appointed by Mr. F. W. Simms, the Resident Engineer, to consist of five, who shall be Masters of workmen, or Contractors of works on the Tunnel, three of whom shall be a quorum ; and who shall meet once a week at least, and shall have the management of the Sick Fund, and regulate the proceedings under the superintendence of the said Mr. F. W. Simms, who is constituted Treasurer of the Fund.

3.—Every man now employed, or to be employed on the works of the Tunnel, shall pay sixpence per week to the Treasurer ; except in case of his having had no more than three days' employment, then he shall not be required to pay any subscription.

4.—Every man must be on the books, and pay two weeks' subscriptions, before he will be entitled to any benefit from the Fund, in the event of bodily sickness ; but he will immediately be entitled to the benefit of the Fund, in the event of personal injury received while in the actual execution of his work.

5.—The allowance to sick members, from the Fund, shall be twelve shillings per week, exclusive of Medical Attendance, in manner following :

A sick member shall be entitled to receive the full allowance for six consecutive weeks ; then if the sickness should continue, he shall be entitled to half-pay for the next following three weeks, when his claim upon the Fund shall cease.

No member shall be entitled to benefit from the Fund, unless his illness or accident is certified in writing by the Medical Attendant ; and if such illness or accident is, in the opinion of the Medical Attendant, or can be proved to be, to the satisfaction of the Committee, occasioned by intemperance, or any other immorality, such member shall forfeit all claim to relief in respect of such illness.

Any member receiving personal injury in the regular course of his employment upon the works of the Tunnel, or in connexion therewith, so as to incapacitate him from attending thereto, such member shall be entitled to full allowance for a period of six weeks ; and then if he is unable from the above cause, to return to his work, such member shall be entitled to half-allowance for a second period of six weeks if his case requires it, when his claim upon the Fund shall cease.

Any member receiving an injury, and being removed to an Hospital, shall, while being an in-patient, receive an allowance of three shillings per week, to pay Hospital fees, &c. : but such allowance shall not exceed twelve weeks. But if he should be an out-patient, and has no other maintenance, he shall receive pay as before mentioned for members receiving personal injuries.

The Committee, with the consent of Mr. Simms, shall have power to alter the foregoing limitations of allowance in any particular case, when circumstances appear to them to require a departure from the general rule.

6.—In the event of a member dying, his representatives, or those entrusted with his funeral, shall be entitled to receive the full amount that such deceased member shall have subscribed to the Fund (exclusive of the sums such member may have received during his illness) in aid towards defraying the expenses of his funeral. Any member leaving his employment, or being discharged, shall have no claim upon the Fund.

7.—The Committee shall have power to increase or diminish the subscriptions of the members, and of reducing the Sick Pay, according to the state of the funds, and the claims thereupon: and shall also make and determine all contracts with the Medical Man for attendance and medicines.

All proceedings and determinations of the Committee relative to the management of this Fund must be reported to Mr. Simms, the Treasurer, and confirmed by him, before they can be acted upon; and when so confirmed shall be final.

The Committee shall prepare, for the information of the members, a Balance Sheet of the state of the affairs of the Fund once a month, or oftener, if they shall be so directed by the Treasurer.

8.—If at the expiration of the works on the Tunnel, or when it may be considered expedient to discontinue this Fund, there should be any funds left in hand, such funds shall be paid for the benefit of the widows and orphans of men who may have lost their life by accident on the works; and those who by accident may have been incapacitated from earning their own living: or otherwise to be given to whatever Hospital or Dispensary for the relief of the Sick Poor that the Committee, with the consent of Mr. Simms, may think proper.

9.—Every Master Workman, Foreman, or Ganger, must give a list of the names of the men employed by or under him, at the Tunnel Office, every Thursday evening by six o'clock, or in default thereof forfeit one shilling for the first hour, and sixpence for the second and every subsequent hour, that elapses after the above time before he so delivers his list, which forfeit is to be the property of the clerk who may have been kept waiting at the office to receive the said list.

Additional Regulations made subsequently to the above.

Any sick member found drinking in a public house shall thenceforward forfeit all claim upon the sick fund, in respect of that illness.

Upon its being ascertained that any member has been, and continues to be, subject to any sickness periodically, or otherwise, the Committee to have power, at their discretion, to return to such member the full amount he may have subscribed to the Fund during the time of his membership, and to declare him to be a member no longer.

HIGHGATE TUNNEL.

It will probably be in the recollection of many persons living, that, early in the present century, an attempt was made to construct a tunnel through the London clay at Highgate Hill, for the purpose of making a more easy communication between Holloway and Finchley. The attempt, however, failed; and the result was the construction of the open cutting, which forms the present Highgate Archway road. The failure appears to have arisen in a great measure from the want of experience on the part of the Engineers who had charge of the work, more especially as they had such very difficult and heavy ground to work in as the London Clay. Those who have witnessed the trouble and difficulty that has been recently experienced in working that treacherous soil will be less surprised at a failure in such a work thirty years ago.

In the year 1811, while the works at Highgate were progressing, the Committee of Management thought it necessary to obtain the opinion of the late John Rennie, Esq., as to the correctness of their mode of proceeding, as difficulties began to appear. That truly eminent Engineer examined and reported upon the works, which report the author has much pleasure in communicating, not only because it will throw some light upon the probable cause of the failure of the work; but also it will dispel the erroneous opinion that too generally prevails, namely, that Mr. Rennie was the Engineer to the said work; whereas the fact was otherwise. The author believes that Mr. Nash, the Architect, was the principal, and a Mr. Vazie, the resident Engineer. It may at the present day be a matter of surprise that an Architect should undertake the construction of a tunnel; but so late as August 17th, 1812, there appeared in the *Star*, a London newspaper, an advertisement from the Regent's Canal Company, addressed to " Architects and Engineers," offering a premium of fifty guineas for the best design for a tunnel that was to be made (and afterwards was made) under the town of Islington; in which advertisement it was stated that the Company were " anxious to have the best information which science and practice can afford on the subject."

MR. RENNIE'S REPORT.

London, Dec. 27th, 1811.

GENTLEMEN,

I examined the Archway at Highgate on Saturday last, and it appears to me there are several parts of the Work which may be altered with much advantage to render the stability more certain, and the works more perfect than they will be if the present mode of execution is adhered to.

First.—On examining the arch already finished, there are two places in which a weakness appears: first, in the upper part of the arch, between the top and side; and in the inverted arch, about three feet from the junction with the side arch.

Second.—The manner in which the bricks are laid, and the bonding of the work together.

Third.—The dimensions of the cast-iron skewplate.

Fourth.—The mode of proceeding to the south, in the progress of the work.

Lastly.—The kind of mortar which ought to be used in the brickwork.

As to the first.—It is quite clear to me that, owing to the swelling of the clay, the whole of the ellipsis is very much compressed; but the upper part, being more curved, is better able to resist that great pressure than any of the other parts; and, therefore, the clay that rests thereon being unable to push it down, slips off a little on each side, and the flat part between the top and sides is thereby charged with this additional load, and yields a little. To prevent this, I advise that in forming the centre this part should be curved about an inch and a half more in proportion than the rest, so that when compressed by the incumbent weight it will come to the form I have drawn it.

In like manner the inverted arch yields, within three or four feet of the iron skewplate. This part should also be increased in its curvation about two inches or two inches and a half, when first built, so that by the compression it will come to what I have drawn it. The inverted arch need not be more than eighteen inches thick in the middle, but it should gradually increase to two and a half bricks as it approaches the skewplates, and continue so to the said plates.

The manner in which the bricks are laid in the arch is by no means calculated to produce the greatest strength; for, as the radius of curvature increases the width of the brickwork at the outside, unless the bricks were radiated they would not bend in the extrados and intrados equally to resist the pressure: the length of the extrados towards the top of the arch is greater than the intrados, by about three inches on every two feet; so much thicker will therefore be the mortar-joints; and mortar, being more compressible than the bricks, when green will yield in proportion. The arch should, therefore, be laid in single rows throughout the whole thickness, until the length in the extrados (or outside) admits of one brick in thickness being inserted there; at which place or places the bricks should be laid lengthways to bind the different rings together: and so continuing round the whole. By this mode the arch will be equally solid throughout, and the immense pressure which is now exerted on the intrados (or inside), so as to flush the bricks, will thereby be avoided: and very great care should be taken

2 A

in making the whole brickwork solid, with as thin joints as it is possible, to make them lie fair in each other.

Third.—The cast-iron skewplates should be made the whole breadth of the joint between the inverted and side arch. This was so represented, in stone, in the section I made.

Fourth.—It would not be advisable to proceed southward with the present arch until the water has been thoroughly drained off. For this purpose, two new pits should be sunk; and the work carried on from these to a junction with the present work.

Lastly.—For nine inches from the extrados (or outside) the cement should be all Roman cement, of the best quality :—from thence to the intrados it should be one part of Roman cement to one of lime, with the proportion of sand. Perhaps Mr. Charles Wyatt, the maker of Roman cement, may know better than me how much lime and sand the Roman cement will bear, as I have not been accustomed to use it mixed.

The extrados, or back of the arch, should have a good coat of Roman cement over it, to prevent the penetration of the water, and between the arch and what is cut out, clay should be rammed as hard as possible, so as to make the action as nearly equal as can be.

In addition to the queries which I understand were put distinctly to me by the Committee, I must beg leave to remark that when I gave the thickness of the brickwork in my section, it was expressly understood that that was the least thickness which ought to be given on the supposition that the clay was perfectly firm and hard where such arch was to be made, but whenever the clay is deficient, and does not answer the description, an additional thickness was to be given, and this at the discretion of the Superintendent, as the nature of the clay should appear—some places will require to be made 2½ bricks, some 3, others 3½, and perhaps some may even require 4 bricks. This, however, can only be ascertained by careful examination; and the Superintendent should be attentive, and decide with judgment; on which decision the success of the work will in a great measure depend.

I am,

GENTLEMEN,

Your most humble Servant,

JOHN RENNIE.

To the Committee of Management
of the Highgate Archway.

ADDITIONAL APPENDIX.

As many varieties in the form of Tunnels, since the First Edition of this Work was issued, have been adopted by different engineers upon the respective lines under their control, to suit the various strata through which they had to pass, it has been thought necessary to give several additional plates to shew some of the figures that have been chosen for the purpose, these with their descriptions form the additional Appendix; their peculiarities will be best understood in the accompanying representations. Plate 13 shews a tunnel entrance on the Wilts, Somerset, and Weymouth Railway for a double line, constructed with brickwork in mortar, having an internal width of 28 feet, with curved retaining side-wing walls. Sections of these are given at several places to shew their construction and relative heights, thickness and batter, and also the projecting footings of the set-offs, as well as the stepping-up of the wing-walls at the tunnel entrance. The whole of the work is finished with a plain weathered stone coping at a height of 29 feet above the line of rails.

The tunnel, which passes through a rocky strata, has a rusticated stone facing projecting outwards from the entrance face, with a batter of 1 in 20; the soffit of the arch being 28 feet above the rails. The crown of the tunnel gradually diminishes inwards the length of 35 feet 9 inches, after which it takes the form represented by the cross section. The points from which the radius of the crown, sides, and invert are struck, are shewn by dotted lines.

The invert, curved side-walls, and arch is $2\frac{1}{2}$ bricks thick, constructed with five $4\frac{1}{2}$-inch rings in mortar, the crown being turned in Roman cement. The ballast on the invert is 2 feet 6 inches deep in the centre, and the horizontal dotted line at the starting of the side-walls is the level of the rails. The thickness of the brickwork at A, on the longitudinal section is shewn in the plan, taken on the same line. Plate 14—The elevation here given is of the front of a

tunnel on the South Wales Railway, having semi-elliptical retaining
wing walls, the entrance being 28 feet in width, and the longitudinal
section taken on the line A of the plan, shews the tunnel entrance. The
brickwork above the centre of the tunnel is 5 feet in thickness, gradually diminishing to 2 feet 3 inches at the termination of each wing-
wall, where it intersects the adjoining slopes of the excavation. The
cross-sections B, C, D, E, and F, shew the relative heights and thicknesses
of the brickwork and batter of the wing-walls; the corresponding
letters of reference on the plan give the various dimensions at the same
places. The parapet wall over the tunnel and wing-walls is weather-
coped with stone 2 feet 3 inches wide and 12 inches thick. The bold
projection surrounding the arch at the entrance is composed of
rough-faced stonework; the key-stone of which is 6 feet 6 inches in
depth, and projects 2 feet 9 inches from the face of the tunnel. The
stones forming each side of the arch from the key-stone gradually
diminish to 5 feet 6 inches in width, resting on a stone base 10 feet
6 inches wide, bedded in brickwork.

The soffit of the arch at the entrance is 28 feet above the level
of the rails, gradually reducing to a point inwards 35 feet 9 inches
from the tunnel face. The cross-section of this tunnel is similar to
that shewn in Plate 13, except the brickwork which is constructed
with six $4\frac{1}{2}$ rings in mortar and the crown turned in cement.

The elevation shewn in Plate 15 represents the entrance of a
tunnel constructed on the Manchester and Leeds Railway and
tunnelled through rock for a single line of rail. The width is
15 feet, and the soffit 17 feet above the level of the rails. The side
curved walls are founded in the rock at a depth of 3 feet 6 inches
below the rails. The rock being of a sound compact nature, enabled
the engineer to form the floor of the tunnel thereon, and consequently
allowed him to dispense with the usual invert.

The centre portion of the tunnel face projects 3 feet before the
curved side retaining wing-walls. The footings of the wing-walls
are stepped up the slopes of the excavation, and are dressed off at
a rate of $\frac{3}{4}$ to 1. The face of the arch is finished with a head and
stretcher chamfered stone block-in course as shewn in the elevation.
The tunnel face is surmounted by two rows of block-in course, each

course, sailing over 6 inches and finished by a stone pediment 20 feet long and 4 feet deep in the centre, and 2 feet 6 inches at the end, with return walls of the same height, 18 inches thick.

The slopes of the excavation are dressed down to $\frac{3}{4}$ to 1, and a puddle french formed below the 6-inch stone pitching at the foot of the slope over the tunnel, to prevent any surface water from gaining access to injure either the brickwork of the arch, or the tunnel face. The pier-points at each end of the tunnel are constructed in Roman cement for about 20 feet.

The cross-section of the tunnel shews the position of the rails and the depth of ballast on the tunnel floor, which has a central drain, as shewn in the section, covered with a selected 4-inch pavement, with open joints to allow any soaking to pass through into the drain. The thickness of the brickwork for the construction of the arch is fully given in the section.

The front elevation of the Birkenhead Tunnel, Plate 16, which is for a single line of railway, is built with chamfered Ashlar facing, having a bold projecting stone cornice, surmounted with a parapet wall, finishing at each end with pilasters. The wing-walls are curved on the face, and run in the direction of the bottom of the slopes of the cutting, behind which are brick inverts laid along the slopes, to receive any surface water and conduct it down to the open channels along the bottom of the slope of the excavation at each end of the tunnel entrances.

The width of the tunnel is 15 feet, and the soffit of the arch is 14 feet 6 inches above the line of rails. A stone channel runs on each side of the tunnel, at the bottom of the curved side walls, level with the top of the rails; these channels act as drains for the tunnel. The ballast is laid 1 foot 6 inches deep on the invert, and the sleepers are embedded in the ballast.

THE END.

THOMAS SCOTT, PRINTER, 1, WARWICK COURT, HOLBORN.

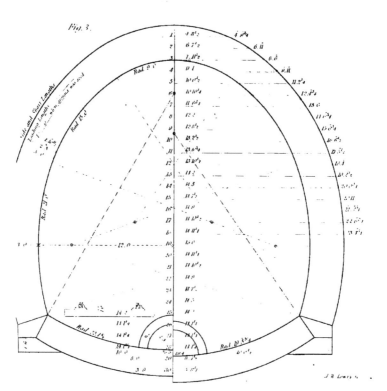

Pat. 1

Transverse Sections.

Fig. 3.

BLECHINGLEY TUNNEL.   SALTWOOD TUNNEL.

J. R. Lewis.

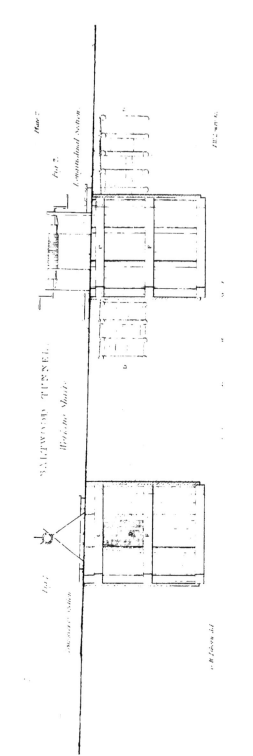

Plate :

Fig 2.

Longitudinal Section

SALTWOOD TUNNEL.

Working Shafts

Plate 3.

Fig. 2.

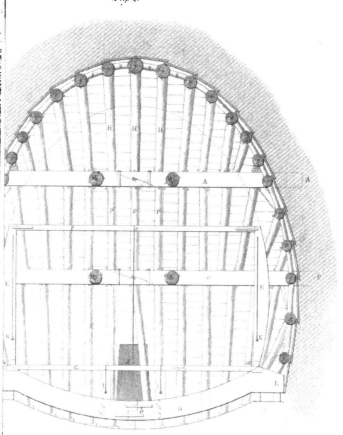

Plate 4

Fig. 2.

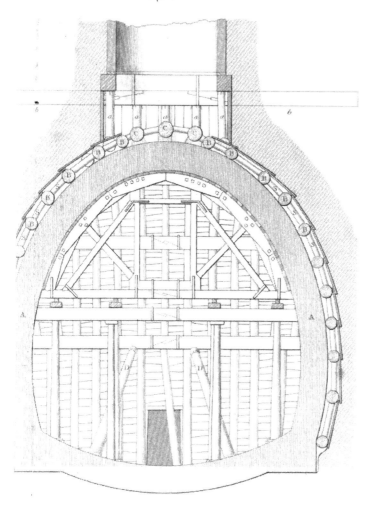

Plate 5

Fig. 2.

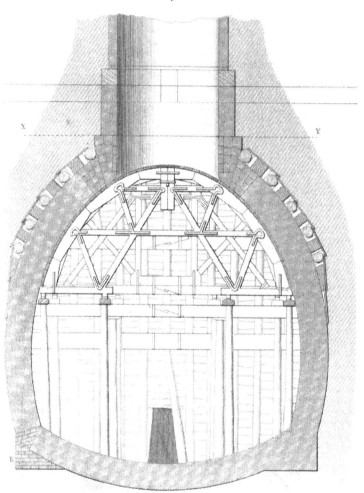

X   Y

Plate 6.

LEY TUNNEL.

THS TIMBERING.

Fig. 2

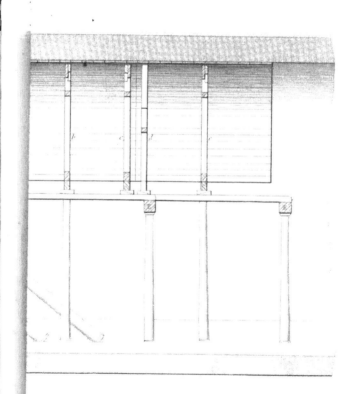

Plate 7.

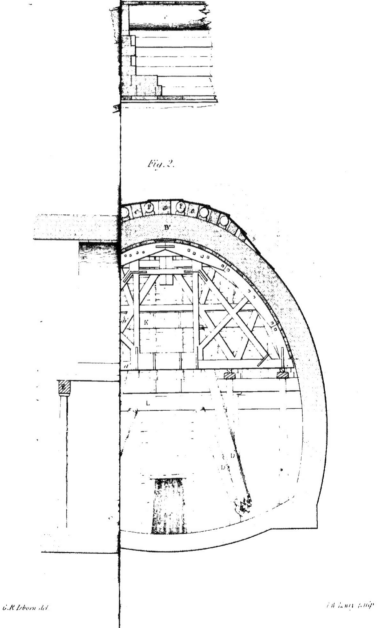

Fig.2.

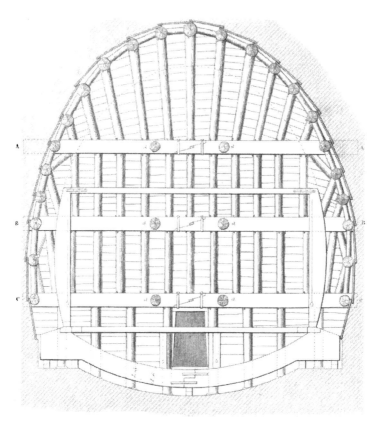

Scale 6 Ft equals 1 In.

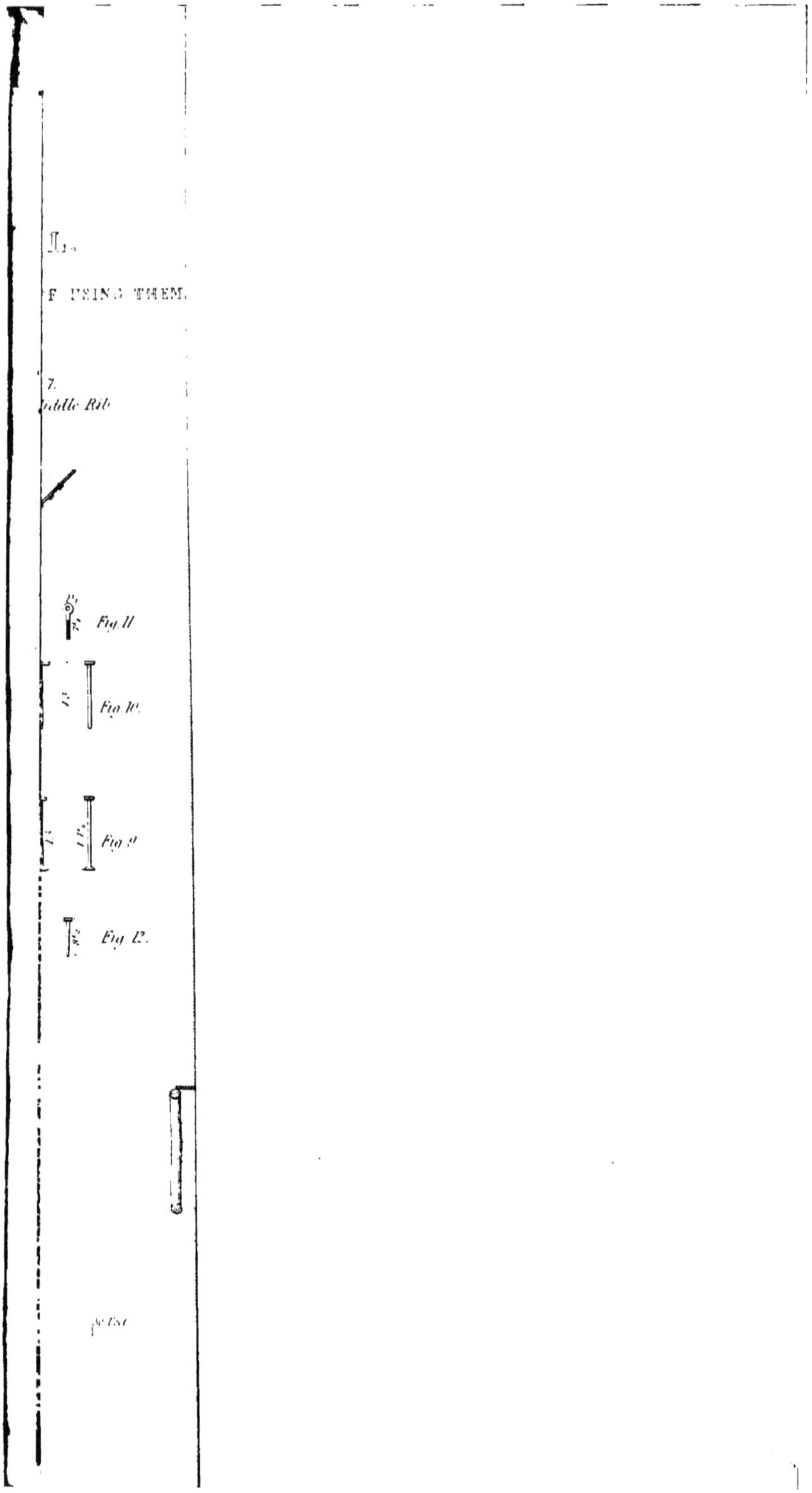

F USING THEM.

7.

iddle Rib

Fig. 11

Fig. 10.

Fig. 9

Fig. 12.

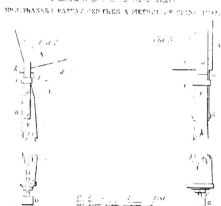

Fig. 1

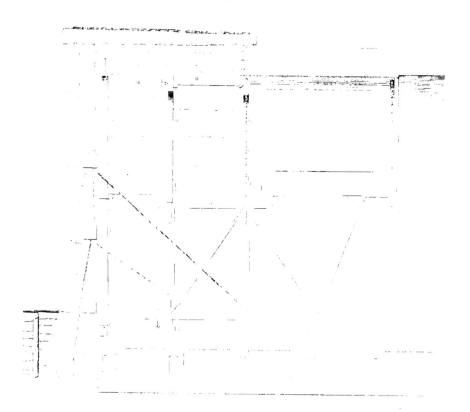

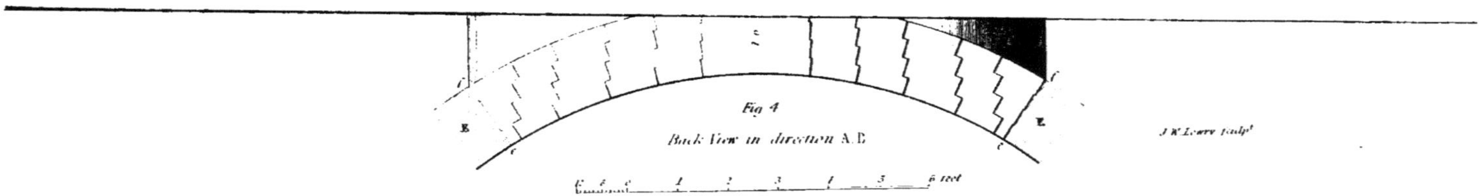

Fig 4.

Back View in direction A.E.

J.W.Lowry sculp.

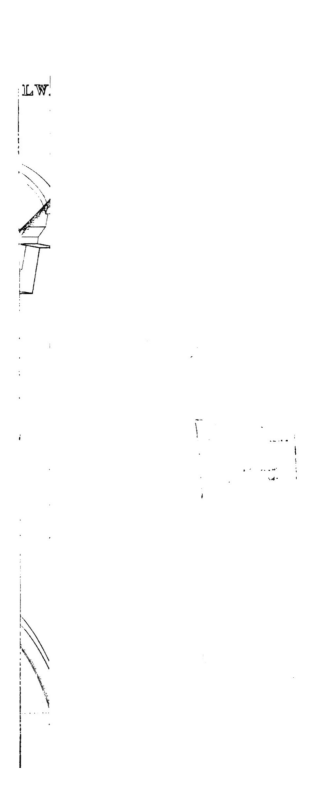

LW.

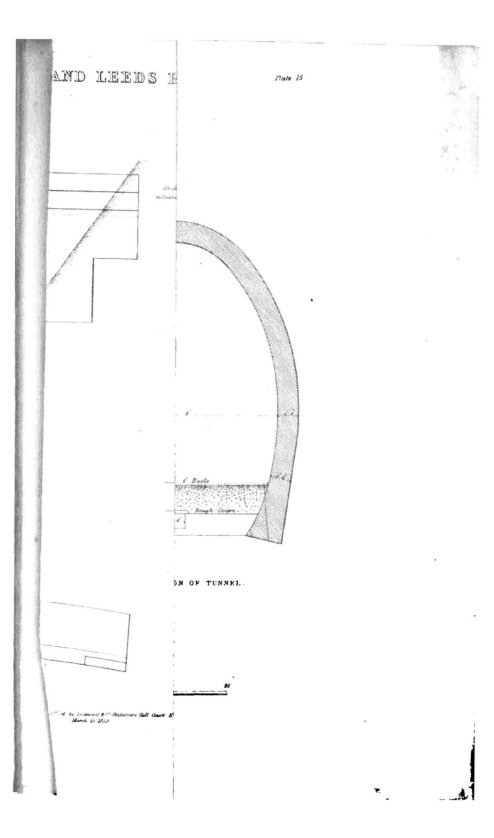

Block
in Courses

6' Rails

Rough Covers

ON OF TUNNEL.

by Lockwood & C Stationers Hall Court E
March 13. 1853

Plate 16

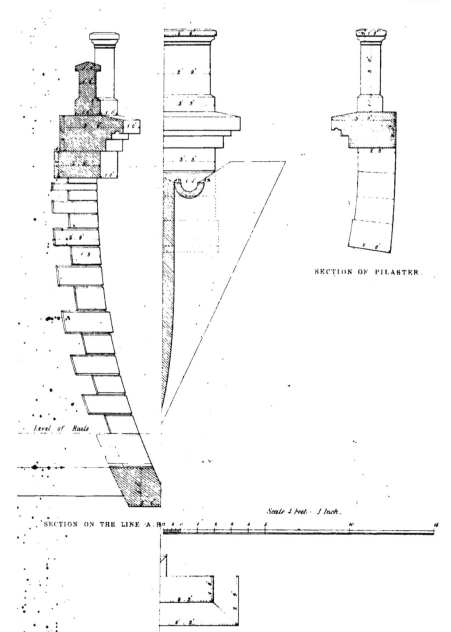

SECTION OF PILASTER.

Level of Rails

Scale 4 Feet · 1 Inch.

SECTION ON THE LINE A.B.

J K Jobbins.

Milton Keynes UK
Ingram Content Group UK Ltd.
UKHW021847250124
436727UK00005B/48